5-24-71

*Utopian
Motherhood:*
NEW TRENDS
IN HUMAN
REPRODUCTION

PREVIOUS BOOKS:

The World of Teilhard de Chardin
HELICON PRESS, 1961

Perspectives in Evolution
HELICON PRESS, 1965

Evolving World; Converging Man
HOLT, RINEHART AND WINSTON, 1970

Utopian Motherhood

NEW TRENDS IN HUMAN REPRODUCTION

Robert T. Francoeur

DOUBLEDAY & COMPANY, INC.
Garden City, New York

ACKNOWLEDGMENTS

In gathering information and insights for this book, I have been very conscious of my deep debt to countless scientists, medical doctors and associates in the academic and research worlds. But I am also very much indebted to reactions and insights on the social and human implications of research into new modes of reproduction which have come so generously from friends and from my students at Fairleigh Dickinson University.

Acknowledgment is also due for quotations of material from the following sources:

The Louis Harris Survey, from *Life* Magazine, June 13, 1969 © 1969 Time Inc.

The Biological Time Bomb by Gordon Rattray Taylor © 1968 New American Library World Publishing Inc.

Evolution and Man's Progress edited by Hoagland & Burhoe © 1962 Columbia University Press.

Man and His Future edited by G. Wolstenholme © 1963 Little, Brown & Co.

The Future of Human Heredity by Frederick Osborn © 1968 Weybright & Talley, Inc.

"The Control of Growth in Plant Cells" by F. C. Steward. Copyright © 1963 by Scientific American, Inc. All rights reserved.

Medical-Moral Newsletter edited and published by Frank J. Ayd, Jr., M.D. © 1968 & 1969.

Excerpts reprinted by permission of The World Publishing Company from *Honest Sex: A Revolutionary Sex Ethics By and For Concerned Christians* by Della & Rustum Roy. An NAL Book. Copyright © 1968 by Kirkridge, Inc.

"Coming Soon: World's First Artificial Baby" by Nino LoBello. Reprinted with permission of Science Digest. © The Hearst Corporation.

"Jesuit Denies Abortion in Early Pregnancies" by James F. Andrews © 1968 The National Catholic Reporter.

CONTENTS

*"To my wife, Anna,
and daughter, Nicole,
who contributed much to the birth of this book.
I hope they and mankind
will find in the world
this book describes
a richer and more human way of life."*

In this book the proposition is set forth that mankind, men and women together, *we* stand on the brink of a major revolution far more serious than that which began a century ago with the emancipation of woman from hearth and nursery. The eternal mystiques of masculine and feminine that have guided Western men and women for centuries are evaporating. Their death is being accelerated by a very short-fused, ready-to-explode biological bomb: mankind's newfound ability to control, manipulate, and direct his own reproductive processes. Techniques of artificial insemination, frozen germ cells, the artificial womb, embryo transplants, prenatal monitoring and manipulations, genetic engineering, and asexual cloning of human beings: these developments, many of them already a reality, will have great psychological, emotional, and religious repercussions. After thousands of years during which reproduction depended on the union of male and female, man stands ready in the next decade or two to reproduce his own image independently.

In separating sexual intercourse and reproduction, man ushers in a new age of crisis and creation for mankind. The human race has demolished its ancient mystiques of male and female by social change and technological developments. But *we have yet to produce their replacements*. Even as the last traces of the old mystiques disappear, men and women today must evolve and create what appears to be totally new images of themselves as sexual persons in a new society. Human sexuality is no longer a biological-oriented reality; it has left that subhuman animal plane to enter a broader, richer and more dangerous world where it becomes a sociopsychological reality involving sexual persons in the depths of interpersonal communications and communions.

Human sexuality is communion-dialogue between persons in their innermost depths as sexual beings. But *we do not yet know what it means to be male or female in this brave new world of ours*. We cannot return to the old mystiques and the morality they engendered, though many will continue to seek security and a role there. We must plunge ahead, with all the risk that entails. We must seek new images, perhaps creating multiple images of man and woman in a new context.

More than any other modern technology, the research into new modes of human reproduction now going on in thousands of biological and medical laboratories around the world is creating a dilemma for man. It strikes at the very heart of human society because it cuts to the quick of the human family and the relationship between husband and wife.

The Creator has somehow shared with us his omnipotence. Having created us in his own image, he now asks us to share with him in the ongoing creation of mankind and man. But we are mere neophytes in the task of creation. We lack wisdom and experience and thus often end up as bumbling, confused gods.

In the pages which follow we offer a triple challenge for the ordinary, everyday man and woman. First, the challenge to acquaint themselves with the background and present state of human reproduction and experimental embryology. Second, a challenge that may provoke them to meditate seriously on the many implications this research holds for them personally as well as for mankind in general. And finally, the challenge to develop wisdom out of this knowledge and meditation so that individual men everywhere may actively participate in the on-going creation of man with some semblance of wisdom.

Our exploding technology of human reproduction brings with it startling new ways in which man may reproduce his image. It brings us to a critical threshold in human history which can thrust us, men and women, into a new and fuller plane of personalization where we become more human, more personal, more male, and more female. But it can also cast us into the asexual impersonal world of assembly-line-produced, thought-controlled, genetically engineered ghosts of men.

The choice is ours. And the pages which follow merely outline that choice in its varied facets.

Robert T. Francoeur
Fairleigh Dickinson University
Madison-Florham, N.J.
September 1970

ONE

SEX GETS
A HELPING HAND
*The Case
for Ethereal
Copulation*

Hope, concern, and a rather sheepish embarrassment marked the
conversation as a wealthy Philadelphian confided his problem to
Dr. William Pancoast. Ten years the senior of his Quaker wife
and a very successful businessman, something was missing in
his married life, a child. Was there, in 1884, anything medical science could do to relieve him and his wife of this burden?

Dr. William Pancoast was certainly the man to approach with
such a problem. After studies at Haverford College and Jefferson
Medical College in Philadelphia with a tour of the medical
facilities in London, Paris, and Vienna, Pancoast had launched a
meteoric career in medicine. In the fall of 1858 he had affiliated
with the Charity Hospital in the city of brotherly love and established a rather large clinic. During the Civil War he served
the Union as Surgeon-in-Chief and second officer of a Philadelphia military hospital. With peace, he returned to private practice and, in 1871, accepted a teaching position in human anatomy
at his alma mater, Jefferson Medical College. With a touch of

nepotism, he succeeded his father two years later as full professor of descriptive and surgical anatomy.

The problem intrigued Pancoast, who brought it up for discussion with the six student doctors in his Jefferson anatomy class. As it clearly appeared to be a case of primary infertility, the lack of fertile sperm or egg, the young doctors first concentrated their attention on the wife. The examination, as Addison Davis Hard, one of the students, later reported, was "very complete, almost as perfect as an army examination," if that meant much in 1884! However, everything pointed to the husband as the cause of the infertility and when he admitted to a youthful bout with gonorrhea, the student doctors tried medical treatment for two months. Finally, in desperation, one of the class suggested a "hired man." Perhaps "the best looking member of the class" might volunteer and during a routine examination of the wife some of his semen could be injected into her womb with a rubber syringe. This of course would be done under anesthetics. Dr. Pancoast approved the ingenious solution and the experiment went ahead without consultation with either the husband or the wife.

Sometime later, after a healthy son, very much resembling the husband, had been born to the couple, Dr. Pancoast began to wonder about the judiciousness of the unusual experiment. He broached the news to the husband and was relieved to find him not at all disturbed about the procedure. His only reservation was to ask Pancoast not to let his wife know how she had conceived.

Professional and personal discretion coupled with the unique and potentially sensational aspects of the case prompted everyone concerned to a strict silence on the matter. When Pancoast retired from Jefferson at the age of fifty-one, his portrait was duly hung in the hall and ignored along with the other venerable professors of days past. His death in 1898 passed unnoticed save by colleagues, friends, and kin. There the matter would have rested except that Addison Davis Hard, a student in Pancoast's class in 1884, brought the whole story into the open in 1909 with an article in *Medical World*. Under the graceful title of "Artificial Impregnation," Dr. Hard related the case and its aftermath, including his trip to New York City where he personally shook hands with the son, who had become a successful businessman, on his twenty-fifth birthday.

Besides recounting what was likely the first documented case of artificial insemination in the human race, Dr. Hard used the experiment as a springboard for some eugenic arguments. This technique, he urged, could be used to improve the human race and eliminate much misery. It was common knowledge among medical men that even the most respectable married women in the city of New York were in danger, since four of every five men in the city were afflicted with some form of venereal disease. Instead of allowing these infected men an opportunity to reproduce any time they so wanted, some serious thought should be given to a new alternative. Venereal disease, Dr. Hard argued, is a plague that produces monstrous offspring. But marriage is a proposition which is not generally subject to good judgment or even common sense. If semen could be collected from respected and successful men, free of the scourge of venereal disease, for use in artificial insemination programs, Dr. Hard claimed men might in one stroke both protect their good women and improve the human race.

Hard's article brought to public notice a major turning point and technological development in the biological revolution that has already placed many aspects of human reproduction and sexuality in the controlling hands of man.

As seems to be common in human affairs, this first announcement of a human being conceived with the help of the physician's syringe substituting for sexual intercourse was greeted with the usual double response: it has been done before and it is against the laws of nature and God.

One letter to the editor of *Medical World* challenged Hard's claim of precedence for Pancoast. J. Marion Sims, who had graduated from Jefferson in 1835, had tried his hand at "ethereal copulation" at least fifty-five times. His percentage of success was low, only about 4 percent, but that was likely due to his belief that the onset of the menstrual flow marked ovulation. (Today artificial insemination is successful in only about 20 percent of the cases.) An Oregon doctor wrote that he had known Pancoast personally as a gentleman who would never stoop to raping a patient under anesthesia. The whole affair, he thought, was the result of a digestive upset on the part of Dr. Hard. Other doctors admitted that the art of insemination was not unknown, but obtaining semen and inseminating with it was either impossible or "ridiculously criminal." This led one doctor to suggest that it would have been much better had the student had intercourse with the patient rather than to break God's law with such a dishonest and immoral procedure.

Generally, though, the reactions were quite favorable. An Arkansas physician reported his success in artificially inseminating mares. If it works with thoroughbred horses, "would not the influence [for good] be many times greater in the human family?" In fact, he suggested, we might encourage castration and reliance on artificial insemination because of the wide prevalence of venereal disease.

In a later article for *Medical World*, dated July of 1909, Dr. Hard admitted that he had somewhat embellished the facts in the first piece in order to trigger some serious thought among both physicians and laymen, particularly on the topic of eugenics. His closing remark was somewhat of a placebo to the Victorian era of Carrie Nation and the God-fearing country he lived in: "I am a firm disciple of impregnation in the good old orthodox manner with all its esthetic features and risk of evil." Even with the confusion of medical precedents and the immorality of tampering with nature, Dr. Hard's report brought home to the public, at least in some small measure and maybe only for the moment, the fact that, without disrespect for either nature or nature's God, doctors then (and much more so today) "modify creation and improve it with intelligence."

Many laymen and not a few scientists and doctors are a bit surprised to learn that artificial insemination was tried successfully on humans by Pancoast shortly after the Civil War, by Sims in the mid-nineteenth century, and likely as early as 1799 by a certain Dr. Home. These experiments give human artificial insemination a respectable, if somewhat thinly drawn early tradition. In surveying the literature of 1909–1910 we find Western civilized man facing, in a roughhewn state, some of the same basic questions we hope to raise with more detail, depth, and sensitivity in this book. In the editorial exchanges of 1909, questions were raised about the monogamous marriage, the right of a couple to unlimited procreation, the desirability of some type of eugenics program to improve the human race, the purpose of sexual intercourse, the meaning of parenthood, man's right to intervene in the procreative process, the role of man and woman, and on through a litany of subsidiary points. Each of these questions is rooted deeply in our understanding of human nature and human sexuality; each draws its nourishment for any answer from the presuppositions and underlying philosophy of man which has for centuries developed and conditioned the whole of our culture. For this reason I am very much convinced that we must be aware of and really appreciate the historical perspective before we can in sincerity attempt any answers to the modern dilemmas of an engineered human reproduction. Before we explore the implications of "ethereal copulation" for this or future generations of man, we must first of all appreciate the process that has conditioned and still conditions our growing awareness of our own nature, our own sexuality, and our own human potential.

That birds lay eggs is obvious to even the most casual observer of nature. Snakes, frogs, and turtles reproduce in a similar way. Even the most primitive of prehistoric men must have noted this simple fact of life, dependent as they were on hunting and gathering food. The sacred books of India reflect an understanding of reproduction that was likely quite common among early men. They divided the animal world into three groups according to their modes of reproduction. There were the egg-laying birds and reptiles where the female was dominant and the male apparently had little or no role in procreation, the higher animals where the seed of the male was nourished in the female's body, and then a large group of lower animals, rats, flies, insects, snakes, and the like, which reproduced without sex by spontaneous generation or from decaying matter.

The spontaneous generation of life, abiogenesis, has held an ancient and honorable position among the explanations of reproduction. The Greeks and other early civilizations accepted it as obviously true. Aristotle, for instance, tells of visiting a pond that had been completely dried up but was later filled with eels and fish spontaneously generated from decaying matter and slime. The Roman poet Virgil records a recipe for producing insects from mud, while other noted scholars report how thunderclouds and rain produced fish and frogs and how honeybees came from the decaying carcasses of horses. In medieval times people believed that worms, flies, and crawling creatures were the spawn of damp putrid matter; serpents were born of women's hair that had fallen into water, and mice could be produced by wheat fermenting in a dark corner with a sweaty shirt. Such views held sway even into the time of our Pilgrim Fathers. It took nearly two centuries of scientific experiments and arguments before this explanation of reproduction was finally refuted. Francesco Redi, an Italian scientist, launched the attack in 1668 and was reinforced by Lazzaro Spallanzani in the late eighteenth century. But it remained for Louis Pasteur in the mid-nineteenth century to finally lay the theory to rest in the graveyard of discarded ideas.

The classic dilemma of priority between the chick and the egg has puzzled men for unnumbered years, but it seems strange that for centuries men trapped and raised birds of various sorts, cleaned and cooked them, without ever asking themselves where the eggs really came from. Not until 1604 did anyone try to trace the origin of the hen's egg. Even then the answer proposed by Hieronymus Fabricius ab Aquapendente was mixed with fable as he missed completely the role sexual intercourse and the rooster played in reproduction. He even repeated with approval the classical tripartite division of animals into those that reproduced by eggs, oviparous; those that reproduced by semen and gave birth to live young, viviparous; and those that came from the spontaneous offspring of decay.

Mammalian and human eggs are indeed practically invisible to the naked eye. The follicle which shelters them during their development in the ovary was not discovered until 1672 by Regnier de Graaf, and then he mistook the follicle for the egg itself. A century and a half of hard research and observation passed before Karl Ernst von Baer correctly identified the mammalian egg in 1820. Add to this the facts that the full consequences of sexual intercourse are not apparent till several months later, and that in humans coitus often occurs with no pregnancy resulting, and it is easy to see how men could make one category of animals which reproduces by eggs alone and another in which the male seed is all-important.

The connection between sexual intercourse, with an equal contribution from both male and female, and the resultant pregnancy is not that obvious when you think of it. Just as an adult finds it difficult if not impossible to recapture the naïve ignorance of his childhood, so we cannot really recapture and appreciate the naïve consciousness of primitive man on this point of reproduction and sexuality. Those working in marriage and youth counseling today, however, often encounter an interesting parallel. Despite our sophistication and sex-saturated culture, despite the prevalence of sex education in our schools, counselors, clergymen, and doctors frequently encounter young and old people, even married couples, who are blissfully unaware of the relationship between sexual intercourse and pregnancy. The sexual instinct may come naturally to man and animals alike, but understanding either the process or its ways is a matter of instruction and/or observation rather than instinct.

An Egyptian papyrus of the Twelfth Dynasty, circa 2500 B.C., contains prescriptions for contraceptives, abortion, and inducing permanent sterility. Early Hindu writers seem to have believed that human pregnancy was caused by the union of the male seed and the menstrual blood since menstruation stops with conception. But such indications are too tenuous for us to assume that primitive man understood in even the vaguest way the idea of sexual fertilization and intercourse. Primitive peoples even today often do not associate intercourse and pregnancy. One eastern Australian tribe believes that baby girls are fashioned by the supernatural powers of the moon and boys by wood lizards. In Queensland the thundergod supposedly forms babies out of swamp mud and inserts them in the mother's womb. Spirit children, no larger than a grain of sand, enter the womb through the mother's navel. Hunting a particular kind of frog, sitting by the fire or leaping over it, cooking a special kind of fish, all can lead to pregnancy.

The Pueblo Indians of New Mexico thought maidens could conceive from a heavy summer shower; the Greeks believed Aphrodite was born of sea foam; the Celtic saint Maedoc was conceived when a star fell into the mouth of his sleeping mother; the founder of the Manchu dynasty was conceived when a maiden ate a red fruit dropped on her lap by a magpie; a pomegranate placed in the bosom of the nymph Nana yielded Attis; and Longfellow records how Winonah was quickened by a western wind and gave birth to Hiawatha. As spring winds drift through the wheat fields of Germany the peasants still whisper that the Corn Mother is coming to bring fertility to both animals and the fields alike. Even Thomas Aquinas has that beautiful canard about the conception of baby girls in the first book of his *Summa Theologica:*

> Woman is misbegotten and defective, for the active force in the male seed tends to the production of a perfect likeness in the masculine sex; while the production of a woman comes from a defect in the active force or from some material indisposition, or even from some external influence, such as a south wind which is moist.

The domestication of animals some ten to twelve thousand years ago and the discovery of farming brought man into a world where his whole life, survival, and comfort depended on the fertility of his flocks and his fields. Small wonder then that man began to find sexuality throughout his world: in the planets, in the stars, in paternal sun and the maternal influences of the moon, in the receptive womb of the earth and the forest gods who made crops grow, the sun to shine, the rain to fall, flocks to multiply, and women to give easy birth. Fertility cults abound in every culture, even the most modern. The ritual mating of the woodland gods, the king and queen of May, the Whitsun bride and groom, find many parallels in the history of man.

Yet throughout all this there has run a thread of mystery and confusion, of misinformation, folklore, and fable that conditions us even today. This background influences our every attempt to approach intelligently the questions now being raised by an exploding technology of reproduction control because it colors the images we have of man and woman in society and in relationship to each other.

Hippocrates, a dwarfish ugly genius from the island of Cos in the Aegean Sea, was a dominant figure in Greek science and medicine about the middle of the fourth century before Christ. One of his experiments in reproduction has echoed down to the present, influencing many moral judgments about the life of the fetus in the womb. Noting that hen's eggs took twenty-one days from the time they were laid till the chick hatched, Hippocrates decided to open a series of eggs, one on each of the twenty-one days. By the end of its first twenty-four hours the embryo chick already has a recognizable shape. For this reason Hippocrates put forth an explanation of reproduction which has since been labeled the theory of preformation. The basic question was, How does an animal such as the chicken come into being? Is it formed totally new from an undifferentiated, shapeless mass supplied by one or both parents, or is there some preformed pattern already present in the male or female seed which simply unfolds and grows larger with time? Hippocrates thought the answer was obvious: the egg contains a preformed pattern for development.

Hippocrates' students carried this idea further, noting that while the chicken reproduced by eggs and its offspring undoubtedly was preformed in the female seed, this did not appear to be the case with mammals and man. Humans did not produce eggs like the bird, so the preformed human cannot be in any egg. The human semen has to be formed in all the parts and organs of the body. After traveling through the blood, these miniature models of parts and organs are gathered into complete preformed patterns in the testicles where they form the male seed.

The great Aristotle has been one of the prime molders of Western man's thought on human sexuality. Aristotle's biological work served as a foundation for his more important work as a philosopher and sage who became in the minds of most subsequent thinkers *the* Master with whom no one dared disagree. It is amazing in reading the moral, social, and political pronouncements of the next two thousand years how often Aristotle's judgment of human nature and sexuality forms the basis for very apodictic statements of the position of woman, her relationship with man, the nature and purpose of sexual intercourse, and the like.

Aristotle agreed with Hippocrates' students about the importance of the male semen, but he was puzzled by the woman's role in reproduction. Nature had obviously created each animal and plant, each organ and structure, for a special, definite purpose. Looking at the world around us, it is easy to learn the purpose or nature of any creature by examining the way it functions, for nature has correlated functions with the creature's purpose and nature. Thus, man is active, full of movement, creative in politics, business, and culture. The male shapes and molds society and the world. Woman, on the other hand, is passive. She stays at home as is her nature. She is matter waiting to be formed and molded by the active male principle. Of course, the active elements are always higher on any scale than the passive forms, and more divine. Man consequently plays the major role in reproduction; the woman is merely the passive incubator of his seed. This appears to be the first time in recorded human history a scientific and philosophical argument was worked out for the natural superiority of the male. Males are larger, stronger, more handsome, etc., etc., etc. In human reproduction the male semen cooks and shapes the menstrual blood into a new human being, while in the birds and reptiles the male semen molds the eggs.

Fifteen hundred years later Albert the Great, Bishop of Ratisbonn and his student, Thomas Aquinas, the "Universal Doctor of the Church," compiled magnificent anthologies weaving knowledge and speculations from the Greek and Christian worlds into harmonious systems of thought. Albert the Great was himself an avid biologist, the author of seven books on plants and twenty-six volumes on animals. But his knowledge of sexuality was really primitive, and in some cases a step backward from what was already known. While he refuted such folklore as the hazel grouse being fertilized by spittle and the male stork detecting adultery in his mate by smell, Albert firmly believed that plants were sexless and that weasels were fertilized through the ear and bore their young through the mouth. He did, however, invent and highly recommend a fine contraceptive device: the heel bone cut from a live weasel and worn around the neck.

Thomas Aquinas did not have the practical inclinations of his teacher and preferred, on the subject of sex, to repeat the sage judgments of his other master, Aristotle. Semen was for Aquinas the active principle of reproduction and the menstrual blood or egg the passive molded substance. He offered some interesting metaphors to explain the whole process of reproduction, suggesting that this process was very similar to the way an artisan creates a bed out of wood with his tools. In other analogies he proposed that the father's semen cooked the unformed uterine blood much as a baker does his dough or that the semen coagulates the uterine blood just as certain substances curdle milk.

One of the prime obstacles to man's quest for knowledge and facts about human reproduction was the almost universal edict against any dissection of the human body. Autopsies were strictly *verboten* and the great physicians of antiquity and the Middle Ages could only learn about human anatomy from their studies of animal cadavers. For instance, Salerno, one of the first universities, dropped the study of human anatomy from its curriculum until the thirteenth century when the city governors finally permitted the medical students to dissect one human cadaver every five years. In 1306 Mondino of Bologna gained infamy for his secret and certainly illegal dissection of a human cadaver. We might like to think of this prejudice as typical of the Dark Ages, but the scandal of New Haven, Connecticut, in the 1830's was the tunnel that was rumored to exist (for the benefit of the medical students) between the Grove Street cemetery and the newly founded Medical School of Yale College.

Toward the end of the fifteenth century knowledge of reproduction began to improve. A military and civil engineer who liked to dabble in flying machines, submarines, canals, and waterworks correctly deciphered the meaning of fossils in the hills of Tuscany. (He was also something of a painter, creating the *Mona Lisa* and the *Last Supper*.) Leonardo da Vinci was the first man we know to take a truly scientific experimental approach to human reproduction. He was the first to make medically accurate and detailed drawings of the dissected male and female reproductive systems, of sexual intercourse, and of the human fetus in the womb. A marginal note beside his drawing of the fetus in the womb notes that "the heart of the child does not beat nor does it breathe because it continually lies in water. If it breathed it would drown, nor has it need to breathe because it is vivified and nourished by the life and food of the mother. . . ." The semen is not produced by the blood in other parts of the body but must be made in the testes, for, as da Vinci observed, castrated men and animals are sterile. He even delved into some primitive sexual psychology, recording experiments on sexual stimulants, the mental and physiological basis of sexual activity, and the involuntary nature of some male sexual activities such as erections. Observations of castrated men and animals led da Vinci to remark that "the testicles increase the animosity and ferocity of the male."

Throughout the Renaissance anyone who studied medicine traveled to one of the Italian universities. Thus some eighty years after da Vinci's death, a young English medical student, William Harvey, went to Padua to be instructed by the most famous physician of the day, Hieronymus Fabricius ab Aquapendente. Returning to England in 1612, Harvey set up a profitable private practice which included as patients Francis Bacon and two kings, James I and Charles I. In 1628 a magnificent and revolutionary tome came from his pen describing in detail, for the first time, the circulation of the blood. A new era for medicine opened up. Fame brought Harvey the leisure of the court physician. Accompanying the king on the royal hunts, he dissected and made drawings and notes on every deer or fox killed by the courtiers. A wealth of factual information on embryology came from this work, and after the Civil War Harvey came forth with a major pronouncement: *Omne animal ex ovo*—All animals reproduce by eggs!

In his masterpiece, *De generatione animalium,* Harvey rejected Aristotle's and Aquinas' theory that new organisms grew out of a mixture of semen and menstrual blood. He likewise disagreed with Descartes and others who thought that new creatures came from the union of male and female seed-elements. The egg was far more vital than the semen which served only to trigger the slumbering egg into its development. Hippocrates' theory of reproduction thus returned to life.

This theory soon appeared confirmed when in 1672 Regnier de Graaf operated on several female rabbits and found in them organs very similar to the ovaries of birds. Under the microscope he found their surface covered with many tiny pimples, apparently the eggs of the rabbit. What de Graaf actually saw was not the eggs but the ovarian follicles which contain the eggs. Unaware of this case of mistaken identity, the ovists were delighted with the new discovery. Harvey's principle was quickly extended to include all living creatures, plant and animal alike: *Omne vivum ex ovo.*

Next on the scene was Marcello Malpighi of Bologna, papal physician and a dedicated microscopist, who spent much of his life studying the development of eggs of various sorts. In vain he tried to achieve artificial fertilization of silkworm eggs by treating them with the seminal fluid of males. In another experiment, however, the microscope and a fertile imagination provided Malpighi with a fantastic discovery. The eggs he studied seemed to contain a minuscule but perfectly formed model of the adult animal. If this be true, then the human and all animals must be preformed in the egg stage. Again the ovists, the defenders of Eve, rejoiced.

The fascination of the ovist theory was poetically summed up by Albrecht von Haller, a Swiss contemporary of Bonnet: "Six thousand years ago God created the earth. At the same time he created the ova of all future plants, animals and men and placed them in the bodies of their first parents. . . . The ovum of two hundred billion human beings was once placed by the creator in the womb of our first mother Eve, in so light and diaphanous a form as to escape notice."

But in the town of Delft, where five years earlier Regnier de Graaf had discovered the ovarian follicle, another Dutchman, Antonij van Leeuwenhoek, was introducing his students at the University of Leyden to the wonders of the microscopic world. Besides being a master of the primitive microscope, Leeuwenhoek was at the same time a man of tremendous curiosity about nature who could inspire in his students the same irresistible urge to explore. One day in 1677 a student brought him a sample of semen from a sick man who suffered excessive nocturnal emissions. What the student had seen in the sample under the microscope seemed to him incredible. The master was equally amazed at what he saw. To friends in the Royal Society of London he wrote:

> I have seen so excessively great a quantity of living animacules [in this semen] that I am astonished by it. I can say without exaggeration that in a bit of matter no larger than a grain of sand more than fifty thousand animacules were present whose shape I can compare with naught better than with our river eel. These animacules move about with uncommon vigor and in some places clustered so thickly together that they formed a single dark mass. After a short time they separated. In fine, these animals astonished my eye as aught else I have ever seen before.

But what were they, these tiny "animacules"? Parasites that caused the man's sickness? Infusoria of a type Leeuwenhoek had studied for years? If that were so, how did they get into the man's semen? Urged by correspondents in London to look into the matter in more detail, Leeuwenhoek ignored his scruples and began collecting samples of semen from some friends. They were all healthy but their semen contained the same little animacules! Experiments with animals followed—the milt of pike and cod, semen from rabbits, dogs, and other mammals, and always the results were the same . . . tiny little tadpoles everywhere! They were very frail and so quick to die that some London scholars suggested they might be the product of decay, an idea Leeuwenhoek stoutly rejected as he was a firm opponent of spontaneous generation. Leeuwenhoek pursued his experiments, linking the tiny animacules to reproduction when he found them inside the tubules of testicles he dissected. All of this research was naturally quite scandalous to his fellow citizens and scientists. Other scientists, of course, had seen these little creatures, but being men of delicacy they had refrained from mentioning their discovery in public. Besides experiments such as this were a breach of morality and against God's laws.

The debate roared on, and gained new fire when another embryologist, Hartsoeker, claimed he saw huddled in the head of a sperm a tiny preformed human being. Preformation might be true, but why let women claim primacy as the ovists suggested? The human being was indeed preformed, but in Adam's loins not Eve's womb. The human egg only provided nourishment much as the womb of mother earth provides nourishment for the seeds of plants. Let us never forget, as Aristotle proved, nature has ruled that the male of the species is dominant.

Counterarguments came in from many sides. The Royal Society surveyed scientists around the world and found that no one had ever observed sperm in the female reproductive tract. Leeuwenhoek refuted this with some conclusive observations based on dissections of animals immediately after copulation. But at the same time another of his experiments provided the ovists with their most potent argument against male dominance and preformation of the sperm. Leeuwenhoek and others had noticed that the small green plant lice which infested roses all seemed to be females, at least during the summer. Detailed observation revealed that these aphids reproduce asexually during the summer when no males can be found. Without mates, the female aphids regularly resort to virgin birth. The egg must then be *the* crucial element and the ovists then were right. Defending the "little man" in the sperm, the proponents of the homunculus theory countered with a truly imaginative theory: warm moist air can contain sperm and these so-called virgin aphids are actually fertilized by sperm floating on the summer night's breezes.

Despite Leeuwenhoek's discovery of the sperm in 1677, debate over the role of the sperm continued on even into the nineteenth century. As N. J. Berrill, author of *Sex and the Nature of Things,* has remarked: "In all this there is a dawning insight into the nature of things, of the meaning of sex and reproduction, but *the understanding has been slow to come,* has advanced by fits and starts and has been hard to hold on to." There was, for instance, the venerable old wives' tale that the first mating influenced all subsequent offspring. Belief in telegony, as this theory is called, is still very common among animal breeders. If a champion bitch by chance mates with a stray mongrel or a prize mare mates by accident for the first time with a broken-down draft horse, she is considered ruined for breeding purposes. Poor blood from that first mating will ruin any later offspring. Theologians argued that if the human is preformed in the sperm, then every sperm must have a human soul . . . and that's pretty incredible! On the other hand, worms, polyps, sponges, crabs, and newts could be dissected or broken up into little pieces and still either regenerate the whole animal from a small part or replace a lost limb, tail, or claw. This seemed to some a good argument that each animal is indeed preformed and life processes simply try to maintain or restore the integrity of the preformed archetype.

The year 1776 brings us back to our original question of artificial insemination and the pioneering work of an Italian priest-biologist, the Abbé Lazzaro Spallanzani. Abbé Spallanzani was a very fine biologist who believed in basing theory on actual observations and experiments. He disliked with a passion the dogmatic claims of some scientists, even those of the stature of Carolus Linnaeus, the Swedish naturalist who worked out a magnificent cataloging process for the animal and plant world, giving each creature two Latin names. Linnaeus had claimed that "in all of nature no impregnation or fertilization of an egg takes place outside the body of the mother." There were obvious examples to the contrary, but the dogma was generally accepted by scientists and laymen alike. At least until 1776 when the skeptical Abbé experimented with some frogs.

The Abbé started off with 165 female frogs already in sexual embrace with their mates. Before they could lay their eggs, he interrupted the goings-on, cut open some of the females, and removed their eggs. Placed in pond water, these eggs soon rotted and died. The couples he had left undisturbed laid eggs which developed normally into tadpoles. The question then was whether the eggs were fertilized before they were expelled from the female or sometime later.

After this gross observation of external fertilization (outside the mother's body), Spallanzani dressed up some male frogs in "little breeches of oilskin." As a control, other males were left *au naturel.* The trousered and untrousered males were then placed with females ready for copulation. Eggs from the un-trousered couples developed into normal tadpoles, while eggs laid by females embracing trousered mates rotted unless some semen from the breeches was spread over them. Carrying his experiment one step further, Spallanzani later removed eggs from a female toad, semen from the seminal vesicles of a male, dipped a pencil in the seminal fluid, and spread it over the eggs which then developed normally in pond water. "Thus I called into life a number of animals, by imitating the means employed by nature."

In 1780 Spallanzani strengthened his claim to his title as Father of the Art of Artificial Insemination by some work with dogs. Confining several dogs in his house, he found one bitch obviously in sexual heat after some twenty days. Collecting semen from a male, he then used a syringe to insert the semen into her womb. Sixty-two days later three pups were born, each bearing a re-semblance to both parents. Not quite certain about the inter-pretation of this experiment (for he was an ovist at heart), Spallanzani then filtered some seminal fluid. The spermless fluid was used for insemination but no pregnancies resulted. If he used just the filtered sperm, he got normal pregnancies.

All of this work should have brought the scientists to a recogni-tion of the role of both egg and sperm in reproduction, but some-how no one seemed to be able to take that single step. The good Abbé, for instance, maintained till his death in 1799 (the year Dr. Home artificially inseminated the first woman) that the sperm is merely a stimulus to the development of the preformed germ in the egg.

For a long time artificial insemination among animal breeders did not become a common occurrence. In certain situations, how-ever, it was used, as for instance when a fourteenth-century Arab from Darfour found himself the proud but frustrated owner of a prize mare. His mare was in heat but the stallion he wanted to mate it with belonged to a neighboring and hostile tribe. According to a report in an ancient medical text he armed himself with a handful of cotton wool saturated with the discharge from the vagina of his mare, approached by stealth the valuable stallion and, having sufficiently excited the animal with the scent of the material he brought, he obtained spermatic fluid from him on the same handful of cotton. Hastening back to his mare, which he had been obliged to leave some distance away, he pushed the whole into her vagina, and obtained by that means a foal.

In 1827 Karl Ernst von Baer discovered and properly identified for the first time the egg of a mammal, in this case of a dog. Pieces of information were slowly falling into place. In 1875 the famous experimental embryologist Oskar Hertwig first observed the actual fertilization process in the embryologists' favorite animal, the spiny purple sea urchin. From then on things began to move more rapidly. In the late 1870's several scientists tried artificial insemination. In the 1890's two problems concerned the experts in animal breeding. First was the fact that many European breeders were getting worse and worse results from their studs; partial sterility was becoming a plague. Repiquet, a French veterinarian, suggested artificial insemination might be helpful as a supplement to normal matings. The other problem was the ancient question of telegony, a question artificial insemination might be able to resolve as it would allow a scientific test of just how much influence, if any, one sire has on the offspring of subsequent matings by other sires. In 1899 E. I. Ivanoff was asked by the chief of the Royal Russian Stud Farm to explore the use of artificial insemination to improve the Russian horses. Ivanoff went on to successful artificial insemination of cattle, sheep, and birds, achieving a better percentage of conceptions than could be had with normal matings. Nearly four hundred Ivanoff-trained veterinarians were at work in Russia prior to World War I. By 1938 Russians had inseminated 120,000 mares, 1,200,000 cattle, and over 15 million sheep.

English dog breeders entered the scene around 1890. Commenting on the English work in 1898, Walter Heape suggested several advantages to artificial insemination. The semen of a single champion could serve several bitches instead of just one, dogs of different breeds where habits or size differences prevented normal mating might be crossed and new species produced, and some very valuable information could be obtained on heredity.

Denmark was the first country to establish a cooperative in which breeders shared their prize animals in a program of artificial insemination in 1936. Then, after a test run in Minnesota, the Extension Service of the New Jersey State College of Agriculture joined with the Holstein breeders of the state to form the first American cooperative in May of 1938.

The technological advances between Spallanzani and Ivanoff moved fairly rapidly when compared with the tedious groping for understanding in the millennia before the seventeenth century. But there remained, at least for those interested in artificial insemination, one major obstacle: semen is very delicate and must be used almost immediately after it is collected. Preserving sperm for any length of time was a longed-for accomplishment that even Spallanzani had considered. In 1776 the latter had observed that when exposed to the freezing cold of winter sperm became motionless but often revived when rewarmed. In 1866 Mantegazza froze sperm at −15°C. and revived them. In 1930 Luyet and Hodapp tried freezing sperm in liquid nitrogen for several days. But always the results were poor and the sperm were damaged by ice crystals forming either inside the germ cells or in the medium in which they were kept.

The breakthrough came in 1949 at London's National Institute for Medical Research. Dr. A. S. Parkes, the key figure in the discovery, later recalled the serendipity experiment. Quite by accident, in a laboratory noted for its precision and exacting concern for detail, some egg albumen and glycerol got mixed up with the test solutions used in freezing sperm samples. Glycerol turned out to be an excellent protective agent so that sperm could be stored in dry ice after slow freezing in a glycerol solution. Initially, storage could be maintained for three months with a 67 percent survival rate. Today bull and other sperm are stored over ten years with no genetic deterioration or change, and with only a slight loss of their vitality and fertilizing capacity.

Research in the use of artificial insemination has proceeded at a fairly rapid pace, at least since the turn of this century, and once the protective role of glycerol was discovered frozen animal sperm very quickly became popular among animal breeders around the world, mainly because it made telegenesis, fatherhood at a distance, an inexpensive and very useful reality. Semen from a prize Scottish ram can be frozen and shipped in a small vial halfway around the world for use, at a minimum cost in an artificial insemination program in a developing nation. Each year in the United States over forty million cows are bred by artificial insemination using frozen sperm. With frozen semen, a champion bull can easily father over fifty thousand prize offspring in a single year—something no sane bull would think of attempting without the helping hand of the scientist!

With humans the use of frozen sperm has been much slower, probably because, as Dr. Parkes observed, there is a "lack of enthusiasm for the idea of abolishing in human reproduction the need for contemporaneous and contiguous action on the part of the two sexes." The same might apply to artificial insemination with fresh sperm from a donor except that desire for a child frequently induces a married couple to overcome their "lack of enthusiasm." The first use of frozen human semen, in 1958 by Drs. Iizuka and Sawada, led to the birth of over two dozen perfectly normal children. At the State University of Iowa, about the same time, Dr. Ralph Bunge and Dr. Jerome Sherman likewise pioneered in the use of frozen human sperm after extensive experiments with animal semen. A couple of years ago, at the University of Michigan, Dr. S. J. Behrman and associates reported twenty-nine successful pregnancies with sperm frozen for over two and a half years. Dr. Behrman's laboratory has led in the development of simpler and safer methods for preserving human semen under the sponsorship of the Ford Foundation and the National Institute of Human Development. While the use of frozen sperm is still relatively limited, it is a technique that is being applied in a growing number of cases.

Several years ago an expert in the field conservatively estimated the number of babies born annually in the United States who were conceived by artificial insemination at about 10,000. Today that figure could easily be doubled, to 20,000 annually. With only 20 percent of the inseminations resulting in pregnancy, this means between 75 and 100 thousand inseminations performed each year by doctors in the United States. This, however, may be grossly understating the popularity of artificial insemination. Samuel G. Kling, author of several texts on medicolegal problems, has estimated artificial insemination to be responsible for more than 10,000 births in New York City alone each year.

In 1950 he estimates 20,000 American children started life as a result of artificial insemination. This figure, he claims, rose to 50,000 in 1955 and to over 100,000 in 1958. What he would estimate today is a moot point. The use of artificial insemination thus far has been pretty well limited to remedying human infertility. In Europe, where the practice is about parallel or perhaps a little ahead of that in the United States, other reasons for insemination are more openly accepted. In Russia, because of the lack of any legal status for the child, only three cases of artificial insemination have been reported and these occurred several years ago.

The use of frozen semen has in the last decade complicated the whole issue of artificial insemination while at the same time making it much more practical and acceptable as a technique in human reproduction. A similar, but in certain ways more serious and extensive impact is probable with our developing techniques for freezing human eggs for later use in reproduction. About two-thirds of all the sterility factors can be traced to the woman. Hence a technique whereby human eggs can be stored for later use would be a major step in remedying sterility. Some indications of a forthcoming breakthrough are already apparent in the work of Dr. Jerome Sherman of Iowa and Dr. James L. Burke at the University of Chicago.

This brief account of man's emerging awareness of the facts of life and his expanding control over human insemination provides a perspective and dimension which is quite essential to any serious attempt to understand the present state of the art and our prospects for the future. In reality, effective techniques for controlling and manipulating the process of insemination are a recent development for which mankind in general has been ill-prepared. Psychologically and emotionally even the more educated portions of our population find it difficult to cope with the reality and implications of artificial insemination. This is evident in the simple fact that the vast majority of married couples who resort to this technique as a remedy for their childless condition find it next to impossible to explain their decision to their child, their families, relatives, and friends. Our religious, ethical, and legal systems are in the same blind situation when it comes to dealing with the new embryology. Very little in our traditions, social, legal or moral, has prepared us for this biological revolution. The questions being raised are questions almost totally beyond the comprehension of traditional patterns of thought simply because they present us with possibilities that clearly forebode a major revolution in human society. When some of these possibilities were raised by such visionaries as the mid-nineteenth-century French novelist Diderot and Aldous Huxley in his *Brave New World,* many people were fascinated. The general reaction, however, was to consign these radical visions to the incredible and hence irrelevant world of science fiction: interesting but completely impossible in terms of man's future.

In this context the roles of the professional philosopher, theologian, sociologist, and psychologist as *the* guardians of human morality and social structures have shifted drastically from a pedigogical pedestal from which these learned men could offer to the masses of mankind clean-cut and neat answers to the questions of where man's future lies. In the past the public looked to its professional scholars and religious leaders for *the* answers. Today the problems we face are so radical and revolutionary that we must return to a much broader base for our insights and guidelines. The religious leader, sociologist, and philosopher of past generations could serve as the depository of our deepest values and traditions. In their years of careful study of our traditions and history they could garner and dispense sage counsel and advice. Today their role is quite different. Instead of dispensing answers they must now *catalyze* men everywhere to reach into the depths of their own being and discover there what it means to be human in today's technological world. The answer to questions about man's basic instincts, maternal, paternal, sexual, and social, can only come from mankind as a whole. Today the professional scholar must look to the public for new and more profound insights into human nature. More often than not this new role is that of a *catalyst,* a question-poser, rather than of an all-knowing oracle. His task is to raise pertinent questions in terms of his knowledge of man's past traditions as revealed by his specialized studies. He must search man's past traditions and cultures for clues and insights which may help us to face the new questions and ask them in an intelligent way. We might go even further and say that our task today is not to ask questions, hoping for a clear and definitive answer, but rather to ask questions provocative of dialogue among men everywhere. In dialogue we perhaps will learn to cope with our expanding technology and direct it in fruitful paths.

This is exactly how I envision my own function in writing this book. After presenting a historical perspective behind different aspects of the new embryology, my hope is to raise some basic questions in terms of our Judaeo-Christian culture and Western view of man. These questions may serve as a catalyst for an open, frank, and unemotional dialogue, an honest exchange of viewpoints among concerned men and women everywhere.

Let us begin this dialogue by posing a very realistic and possible application of the new embryology. The catalyst for our discussion is simply an application of the technique for freezing human eggs and sperm for later use in artificial insemination programs. Only two relatively minor problems remain to be solved before this application becomes imminent as a common and important reality of man's daily life. Facilities for the long-term storage of germ cells must be developed in the majority of hospitals around the country and our technology must reach the point where we have the medical certainty that eggs and sperm can be cold-stored, without significant deterioration, for a period of about twenty-five years. The first problem will be solved as soon as public demands convince hospital administrators that certain monies should be set aside for the necessary equipment and for the training of doctors in this technique; the second involves only a minor extension of the research which within a decade has brought us to the point where successful fertilization can be accomplished with sperm that has been stored over ten years.

We might propose that shortly after reaching puberty every young lady should visit her family doctor for a thorough checkup. If everything proves normal, the doctor can give her an injection of hormones combined from pregnant women and pregnant mares. These will shock the girl's physiological system so that instead of releasing a single egg in the middle of her next monthly cycle she will superovulate several dozen eggs. As they pass down her fallopian tubes on their way to the womb these eggs can then be collected with a simple syringe technique for storage in a deepfreeze. The plastic thimble-sized container, of course, will be properly labeled and officially sealed. The girl then can have an operation similar to an appendectomy in which her fallopian tubes are cut and tied. In the future she will have a normal monthly cycle with normal ovulation but since the egg and sperm can no longer come together conception will be impossible.

A teen-age boy might follow a similar procedure, without the hormone injection, and have a large sample of his semen collected for storage. Unlike the parallel operation in the female which involves complete anesthesia and major surgery, cutting and tying the *vas deferens* which carry the sperm out of the testes is a minor operation involving only a visit to the doctor's office and a local anesthetic.

Some years later, these two young people might meet, fall in love, and marry. When they decide to enrich their marriage with a child, they can visit the local dispensary where their eggs, semen, or both are stored. If only the husband's semen was stored and the wife did not have an operation, simple artificial insemination could be used; if only the wife's eggs were stored, the doctor could defrost one of these and implant it back into her tube for fertilization in the course of normal intercourse. If both germ cells were frozen, test-tube fertilization could be followed by implanting of the fertilized egg in the wife's womb.

This suggestion has many promising aspects and very few, if any disadvantages.

A prime advantage is that whether the vasectomy (cutting the tubes) occurs shortly after puberty or during the engagement, the whole controversial field of contraception is bypassed. Mechanical and chemical contraceptives, diaphragms, condoms, spermacidal jellies and foams, are aesthetically and/or morally unacceptable to many people. Very often they interfere with the spontaneity and enjoyment of the sexual communion. Less objectionable are the intrauterine coils and spirals, the IUD's, but their use is frequently limited to women who have borne one or more children. Women who have not given birth for some reason find it hard to retain the small plastic device in their wombs. Some theologians and medical doctors disapprove of the IUD's because they seem to prevent implantation of the developing cell mass, morula, some four to five days after conception. They might not object to a contraceptive, but they find the IUD's tantamount to an early abortion. The now famous contraceptive pills have received much publicity concerning their harmful side effects. Whatever is the final medical decision on their use, it is

quite clear today that they are sometimes associated with an increase in blood clotting, chromosome breakage and duplications
as well as lesser but still disturbing side effects. Furthermore, the
long-range effects of a continual disruption of the female hormonal cycle with a "continuous pregnancy" such as the pill produces are still unknown and more often unsuspected. Conven-
tional contraceptives still in the developmental process do not
seem to hold much hope for really bypassing the same disadvantages we find with those already available. They may reduce
the disadvantages to a somewhat safer level, but since they are
based on the same medical principles it appears unlikely that
they can eliminate them. Whether it be a "morning-after pill," a
time capsule implanted beneath the skin of the woman, or a
monthly pill that arrests sperm production, we are still dealing
with hormones and disrupting very sensitive physiological cycles.
Even the use of antisperm serum that would render the wife immune to her husband's sperm poses medical problems. The procedure suggested above can in fact show up all the contraceptives
now used or known to be in development for what they really
are, very crude and primitive means of achieving responsible
parenthood and a mature, fully human family.

Another important advantage of this procedure is that it could
completely eliminate the whole problem of unwed mothers and
illegitimate children. Several states, including Connecticut, have
published Department of Health surveys indicating that one out
of every six girls between the ages of thirteen and nineteen is
illegitimately pregnant. This is no small percentage! Of course, it
would also expand the problem of premarital sexual intercourse
far beyond what has already been raised by the extensive use of
the "pill" by college and even high school girls. By removing
completely the fear of pregnancy as a repressant, it might also
allow young people to situate sexual intercourse in its proper perspective as a mode of interpersonal communion and communication so sacred and so deeply personal that promiscuity and
bed-hopping with the philosophy of "let's make love but not get
personally involved with each other" would be instinctively
ruled out. If such a system were adopted by the public, it would
require a profound and penetrating analysis of human sexuality
and the interpersonal relationship of men and women on the part
of elders and parents along with a wholehearted effort at in-

depth education of the young from the elementary grades on up. And here I am not thinking of the biological reductionism of human sexuality to sexual intercourse and reproduction which characterizes much of our so-called sex education in American schools *and* homes. Education in the biological facts of human reproduction is only one element in the complete education of a person to his sexual nature. Sex education begins with birth, as soon as an infant begins to respond to the cultural impact of society and his family, composed as it is of sexual human beings. Sex education goes on at every moment of our lives.

Another advantage of this proposal might be the elimination of that tension-provoking period in a woman's life, just before menopause, when her childbearing days are almost over and she is tempted not to bother with contraceptives even though there is an outside chance of her ovulating a fertile egg and becoming pregnant. It might also help a younger woman faced with the psychological and emotional crisis of becoming sterile as a result of an operation for ovarian cysts or tumors.

On the negative side this proposal raises some serious questions about the rights of the individual person. It raises the specter of a further alienation of man from himself and his world in the extension of impersonalism into the heretofore sacred and private realm of human reproduction. The proposal, if implemented, might seriously affect the emotions of men and women by taking the unexpected element, the romanticism and love out of the conception of a child.

Another potential danger lies in the risk the general population would run, particularly in countries where dictatorships abound, that the government might step in to control who reproduces and/or how many children a couple may have. Anyone participating in such a program would have to make an act of faith in his government's integrity and goodwill. A similar act of faith in the doctors would be necessary because of the possibility that they might mix up the sperm or eggs or withhold them when requested. Controls might even be applied in a democracy should the population reach a critical level. It could reach the science-fiction point where a government would set a certain quota for conceptions based on the number of deaths in the population. In this case a married couple might have to wait for the death of some near relative before they could be licensed to have a child.

In the near future major governments may find it necessary to sign population limitation treaties even as we now agree to arms limitations.

Difficulties can arise in the distribution of frozen semen. A woman might place her eggs in storage as a teen-ager in London and years later want to use them in South America. Shipping frozen sperm is a simple matter, but there would have to be some real guarantee that the semen and egg used for insemination actually came from the husband and wife.

We can simplify this proposal in a way which can definitely be applied today simply by expanding our techniques and facilities for sperm banking. This would entail the storage of semen samples from all men as they become teen-agers. The same advantages might be achieved, but in the process we would have to explore certain potential risks. Sterilization, without recourse to frozen sperm banking, has not been accepted in many developing nations of Asia and South America where *machismo*, the male ego bolstered by the number of offspring, dominates the culture. Furthermore, vasectomies occasionally bring on psychological impotency in men.

The situation in Russia today raises some interesting legal aspects for any program of artificial insemination. At present Soviet law does not confer any recognition or status on children born as a result of such insemination. There is nothing in Soviet law which would prevent a Russian husband from disowning such a child even if he had given his permission. Doctors naturally are reluctant to help a childless couple with artificial insemination, despite the fact that fully 20 percent of Russian couples are involuntarily childless, half of them because of male sterility. (In the United States the figure is roughly 15 percent with the male being sterile in 35 to 45 percent of these cases.)

In late May of 1969 the Soviet weekly *Literaturnaya Gazeta* carried a weighty plea from an eminent group of scientists, doctors, and psychologists urging the Soviet Union to allow artificial insemination *for both married and unmarried women* who want children. "There is nothing immoral in the fact that wives or unmarried women, moved by a natural desire to have children, turn to science . . . It is well known that the desire for children is a welcome and strong link in marital union and that unmarried women cease feeling isolated when they have children," the article argued. While American doctors generally limit their use of artificial insemination to married women, the procedure has occasionally benefited unmarried women in Denmark and Sweden. As the Soviet article pointed out: "In our view, artificial insemination is as moral and humane as adopting a child from an orphanage. For many, many families, the help that science could offer would be a blessing. Artificial insemination is as safe and as common, for instance, as a blood transfusion."

If you had a daughter or a sister who had chosen a career instead of marriage or was unmarried simply because she never met the "right man," how would you react to her desire to have a child by artificial insemination? Would this be that much different from her going to an adoption agency where even as an unmarried woman she could, in some states, legally adopt a baby? The results of the two approaches are not that different, even though her own psychological and emotional reasons might be very different in the two situations. And obviously, most prospective grandparents would react quite differently to the two solutions. Several sociological surveys of young people in Europe and the United States indicate a growing trend toward and acceptance of "bachelor mothers." It is questionable whether many parents over thirty would accept the single mother.

Owen Garrigan, a biochemist and priest teaching at Seton Hall University, has raised a parallel question that can nourish some lively theological debates. In his book *Man's Intervention in Nature,* Garrigan asks whether celibate priests and monks should be allowed to make contributions to sperm banks. Would this be compatible with the dedication and the spirit of celibacy and chastity? "Provided that viable sperm could be obtained in some morally neutral and acceptable way, the storing of sperm to be used after the death of the celibate would seem comparable to the donation of blood for a transfusion." Should a nun be allowed by church law or her community to donate an egg to be fertilized by her brother-in-law and carried to term by her sterile and childless sister? A similar case might arise where the husband and wife are fertile but the wife cannot carry the child. Should the nun accept her sister's request that she carry a child conceived from germ cells of the husband and wife? For the monk or celibate priest, a contribution to the local sperm bank would be a minor and momentary thing. It could hardly be viewed as a com-

promising of his dedication, or even a violation of the strict interpretation of his vows. Nevertheless, the Roman Catholic
Church has for centuries laid great emphasis on biological integrity in dealing with virginity and chastity. As Garrigan concluded,

> it would seem that being a donor in the VCOGP scheme [for the
> *voluntary choice of germ plasm* and artificial insemination proposed
> by Hermann J. Muller] is incompatible with the state of celibacy.
> The donation of sperm is not merely a biological event. As an act
> of the person, its value under normal circumstances as a contribution
> to the presumed improvement of the race would not seem to out-
> balance its incompatibility with the dedicated life of the celibate.
> In an analogous situation, it would evidently be quite disruptive of
> the present order and an apparent violation of chastity if religious
> women offered themselves as "foster mothers" for embryos trans-
> planted to their wombs from women who had conceived, but could
> not bear, the children.

Would you agree, or disagree?

These possibilities may upset some people, but they remind
me of a sly question posed during my oral examination for the
doctorate in biology. Dr. Arnold Clark, a geneticist at the University of Delaware, asked if I thought the Roman Catholic
Church was really antievolutionary. By forbidding priests to
marry during the Middle Ages when the elite of Europe were
either nobles or clerics, the Church drastically reduced the percentage of desirable genes in the gene pool. If the Church were
really interested in the good of mankind, Dr. Clark suggested, it
would urge priests to marry so that their genes for intelligence,
good looks, etc. might become an active contribution to the gene
pool and spread throughout the human race. The other professors
on the examining board chuckled, but it is an interesting question. Owen Garrigan's proposal of artificial insemination with
frozen sperm from clerics might offer a way around the dilemma.

In some states the courts have made it acceptable and legal
for an unmarried woman or man, even a homosexual, to adopt a
child. This appears to be a trend in both American and Conti-
nental cultures so that perhaps our next generation may find
bachelor mothers, either by artificial insemination or by natural
means, not at all out of place in society. In Sweden unwed

mothers are anything but uncommon. They are legally entitled to use Mrs. as a title. Nor are they excluded from any of the family allowances and benefits provided by the state for married mothers. Their offspring are considered legitimate. In the United States the legal situation of unwed mothers is very clear, but the question is whether the presumptions of the law that the child was conceived in the natural way would carry over to an unmarried woman who chose to have a child by artificial insemination. This question may become quite serious if the present near balance of males and females is drastically upset because of war or some other cause. If the number of women in the twenty to thirty-five age bracket moves far ahead of the men available for marriage and family life, perhaps unmarried mothers might become socially and legally more acceptable.

Artificial insemination has been almost completely ignored by the legal apparatus and bar associations. Individual cases arise here and there with very little guiding precedent for the judges or juries. Only the state of Oklahoma has given status to artificial insemination by legalizing it in 1967.

Two facts are very clear, however, in the American legal code. Adultery is a criminal act in forty-five out of the fifty states and every state accepts proof of adultery as grounds for an absolute divorce. In this context artificial insemination poses some very serious legal issues. When artificial insemination with a donor's semen is used, the husband is in no way the genetic or biological father of the child. The child's real father is *not* the husband of the mother. Does this boil down to a case of modified adultery? If it does, then the child is illegitimate. Furthermore, a consenting husband runs the risk of criminal conspiracy, for no one can legally consent to a criminal act. In 1921 the Supreme Court of Ontario, Canada, ruled that artificial insemination with a donor's semen [AID] is adultery if the husband does not consent to it. The judge explained his decision and its implications rather ambiguously when he stated that "if it was necessary to do so, I would hold that [AID] in itself was 'sexual intercourse.'" In a 1954 case an Illinois judge went further and declared that artificial insemination was adultery whether or not the husband gave his consent.

On the other hand, in an unreported case in Illinois Judge Feinberg held that AID is insufficient evidence of adultery to

serve as grounds for a divorce decree. A 1911 New Jersey Court
pointed out that adultery must include some actual bodily union
of the two partners. Thus, as a Pennsylvania Court concurred in
1943, artificial insemination cannot legally be considered adultery. A decade ago, a subcommittee of the Michigan State Bar
Committee added its weight to this appraisal as did a famous
Scottish case presided over by Lord Wheatley in 1958.

Typical of the legal contradictions is that prevailing in the
state of New York. The State Supreme Court ruled in 1948 that
children conceived by artificial insemination are to be considered
legitimate when the husband has consented to the procedure.
Yet the City of New York has on its statutes a law declaring illegitimate those children born to a married woman which are "begotten at a time when the husband is impotent." New Yorkers
are thus in a peculiar position. If a husband is "impotent" because of a very low sperm count, he might have a physician
concentrate several batches of his semen for use in artificially
inseminating his wife. He then faces the possibility that the New
York City Criminal Courts might subsequently declare the child
illegitimate since it was conceived at a time when he was tech-
nically impotent. On the other hand, a principle of law presumes
the legitimacy of all children conceived in wedlock, and all children so conceived are presumed to have the same legal rights.

If the courts of New York City consider artificial insemination
a form of adultery, it is the only form of crime for which a procedure has been carefully specified by law. In the Sanitary Code
of New York City for 1947 and 1950 we find the following
regulations:

REGULATION 1. A person from whom seminal fluid is to be collected for the purpose of artificial human insemination shall have a
complete physical examination with particular attention to the
genitalia at the time of the taking of such seminal fluid.

REGULATION 2. Such person shall have a standard serological
test for syphilis and a smear and culture for gonorrhea not less than
one week before such seminal fluid is obtained.

REGULATION 3. No person suffering from any venereal disease, tuberculosis or infection with brucella organism, shall be used as a
donor of seminal fluid for the purpose of artificial insemination.

REGULATION 4. No person having any disease or defect known

to be transmissible by the genes shall be used as a donor of seminal fluid for the purpose of artificial insemination.

REGULATION 5. Before artificial human insemination is undertaken, both the proposed donor and the proposed recipient shall have their bloods tested with respect to the Rh factor at a laboratory approved for serology by the Board or Commissioner of Health. If the proposed recipient is negative for the Rh factor, no semen shall be used for artificial insemination other than from a donor of seminal fluid whose blood is also negative from this factor.

REGULATION 6. Where artificial human insemination is performed, the physician performing the same shall keep a record which shall show:

1. The name of the physician.
2. The name and address of the donor.
3. The name and address of the recipient.
4. The results of the physical examination and the results of the serological examination, including the tests of the Rh factor.
5. The date of the artificial insemination.

Such records shall be regarded as confidential and shall not be open to inspection by the public or by any other person than the Commissioner of Health, an authorized representative of the Department of Health and such other persons as may be authorized by law to inspect such records. The custodian of any such records, the said Commissioner or any other person authorized by law to inspect such records shall not divulge any part of such records so as to disclose the identity of the persons to whom they relate except as provided by law.

Can a husband cite the donor as correspondent in a divorce case? Can the wife be tried in criminal court for adultery and sentenced to five years in prison if she is a resident of Connecticut, or a thousand-dollar fine if she lives in Maine? Is the anonymous donor guilty of adultery in either the moral or legal sense? Some years ago a New York doctor sought a divorce from his wife on just these grounds. At the time adultery was one of the few grounds for divorce in New York State, and the husband who had agreed to artificial insemination for his wife used this fact as the equivalent of adultery in a court case. In this instance the courts did not agree with the husband and the divorce was denied.

Such precedents place judges and juries in future cases on the horns of some pretty agonizing dilemmas. For instance, if AID constitutes adultery, is the anonymous donor the adulterer? This is exactly what one irate physician charged when Dr. Pancoast's pioneering insemination became public in 1909! Or do we lay the moral stigma of adultery on the physician, since he obviously is the one primarily responsible, even if by way of a rubber syringe, for the impregnation? The Gordian knot becomes even more tangled when the gynecologist who does the insemination is a woman.

Picture, for instance, the case of a young scientist who decides to store some of his semen because of radiation dangers at his place of employment. Later on he and his wife have several children by artificial insemination. Meanwhile, some unscrupulous lab technician decides to pick up a few extra dollars by selling the leftover semen of this brilliant scientist to a second couple. The bootlegged semen is used for AID. The second woman's husband later takes her to court for adultery and names the unsuspecting scientist as correspondent in the divorce proceeding as well as asking that he be liable for support of the child he did not even know he had fathered!

Every year in the United States some ten to twenty thousand babies are introduced into a legal no-man's land. Take the case of a married couple who agree mutually to have a child by artificial insemination using an anonymous donor's sperm. If the couple later are divorced in the state of New York and the wife is awarded custody of the child, she cannot legally deny her former husband visiting privileges on the grounds that it is not his child. However, if she moved to another state, as one wife did in this situation, she might find a legal precedent in court. Despite the fact that the husband gave his consent to the insemination, loved and nourished the child for several years as his own, some state courts have ruled that he is not the father in any sense and therefore not entitled to visiting privileges. This contradiction highlights the ridiculous positions judges and courts often end up in when suddenly confronted with the results of technological developments for which the law has no clear precedent.

We can carry this quandry several steps further and ask whether legally a separated husband is bound to support a child conceived by artificial insemination until its twenty-first birthday? In 1967 a man divorced three years earlier was charged by the District Attorney of Sonoma County, California, with willful nonsupport of his wife's six-year-old child conceived with his consent by AID. The wife had refused such support, but the man was found guilty. Even though he may have agreed to the venture, can such a husband disown the child as illegitimate when born? If the husband is not "in any sense the child's real father," according to the courts, then who is the father of the child? The anonymous donor? If so, could the donor claim the child, or at least visiting rights? Can the child claim a share in his biological father's estate? If the identity of the donor becomes known can the husband sue him for child support? One way around this problem is for the husband to legally adopt his offspring by AID.

While we are exploring the legality and morality of artificial insemination, it might be worth raising the specter of incest. With frozen sperm and artificial insemination, it is quite conceivable that a young lady might want to have a child by her father or grandfather or uncle who happened to be a judge on the Supreme Court, a Nobel Prize scientist, or a great composer or artist. This would fit into much the same category as the question of adultery, with perhaps some added gravity, because of the more serious character of incest. Many will agree with Colin Clark, the respected English biologist, that "artificial insemination by a donor other than the husband has all the malice of adultery [or incest]—it is a form of adultery—and I think that anyone who understands the moral meaning of the word adultery is bound to reach that conclusion." Despite Dr. Clark's sureness, I still doubt that all my readers would agree with him.

Artificial insemination, especially with frozen semen, poses many moral and social questions. How do we record on a birth certificate the fact that a child's father died some ten years before he was conceived? Or do we record that fact at all? What kinds of prejudice do we expose a child to in this situation? Would a widow be given that strange look neighbors occasionally cast about if she had a frozen sperm baby by her deceased husband? Would the child be considered illegitimate, in practice or even in theory? How do we deal emotionally, socially, and legally with telegenesis, the ability of a man to become a father halfway around the world or from the moon by shipping his frozen semen for use in artificial insemination? What happens to our concept of fatherhood in telegenesis? What is likely to happen to our family structures with this development already a reality?

Let me complicate the situation a bit more by adding yet another dimension that weaves through the emotional, psychological, legal, and medical aspects of our life. Everyone is aware of the fact that man has been intervening in the breeding of animals for centuries, selecting parents on the basis of specific desirable traits and then breeding only certain individuals in a population. For centuries dog and cat fanciers have created substantial changes, ranging from the Mexican hairless to the Pekingese, from the dachshund to the grayhound. Man has stepped in to direct the mating of pigeons, chickens, cows, guppies, and goldfish along what he thought were either desirable or just interestingly bizarre lines. In some areas these experiments have met great success; in others, like the increased viability and fertility of livestock, not so. Since the turn of the century when Francis Galton first proposed a scientific program of selective reproduction in the human race, man has become increasingly aware of the possibilities of improving the race by eugenics.

The very thought of eugenics is a jarring irritant to many people, particularly in the light of the Nazi attempt to produce a superrace of men. But men have indirectly engaged in eugenics for many years. Plato argued that in the Republic the right to reproduce should be limited for the common good to men selected for their desirable qualities and intelligence. For the better part of human history, for over a million years, the human race survived with a polygamous society. Under this system a king or important personage in the community was allowed to gather around himself many wives and to sire many more offspring than the less favored or less intelligent members of the population. King Solomon, for instance, had the opportunity to pass on his proverbial wisdom and intelligence to the offspring of his six hundred wives and over three hundred concubines! In a polygamous society where leaders are selected for their intelligence and acumen, as well as other desirable traits that lead to success in human affairs, we should expect that the process of improving the gene pool and man's heredity should go on at a much faster rate than is possible in a monogamous culture. There is no way, though, by which we can test the effects of this early experiment in eugenics and selective reproduction. Despite the ideal condition of monogamy which has been the goal of the Judaeo-Christian world, man's attempts toward a strictly monogamous condition have not in any way been a total success. Mankind has never *really* crossed the biological hurdle from polygamy to monogamy. Thus a strict comparison would be weakened even if we had plenty of data on the offspring in polygamous and monogamous cultures.

J. B. S. Haldane, the delightfully eccentric British biologist, has found another precedent for the selection of donors in an "artificial insemination" program in an Indian tradition. Until recently, the Jats of northern India integrated a form of eugenics into their social structure which includes ordinary fraternal polyandry.

A young man judged of outstanding merit for physique, courage, and other qualities, was allowed access to all married women of a village. He was given a pair of gilded shoes which he left outside the door when performing his eugenic duties, to warn off an ordinary husband. After fifteen years or so, when his daughters became nubile, he was killed to avoid inbreeding. But he might, and often did, leave the village with a chosen partner. Having fought in the same brigade as the 6th Jats, I can testify to their courage and efficiency as soldiers. In view of such traditions, the choice of a father other than a woman's legal husband may arouse less opposition in some parts of India than in other countries, whether artificial insemination or the normal process is employed.

Unlike the Judaeo-Christian tradition, existing Hindu laws and morality permit this practice, for which Julian Huxley coined the appropriate term, "pre-adoption." Instead of selecting the child, you select his father.

Hermann J. Muller, who won the Nobel Prize in the 1930's for first inducing hereditary changes by irradiation, was for years a staunch and vocal advocate of eugenics. Even when he went to what most people would judge an extreme in his arguments, Muller often touched on some very profound points. For instance, he has asked us to face a question few of us ever think of facing, namely, how much personal conceit and selfishness is there in our often vehement attachment to having our own children in the traditional sense. As an example, why do some childless married couples spend thousands of dollars trying to conceive a child of their own when there are thousands of orphans and abandoned children waiting for adoption? Why does a couple go through all the examinations and preparations for having a child by artificial insemination with a donor's semen, when they can adopt an orphan or abandoned child who will otherwise never know the joys of a loving family? In some cases, when a couple is for various reasons eager to adopt a racially pure child, artificial insemination with a donor's sperm can prove to be an easier solution than the complications and delays of waiting for an unwed Caucasian girl of proper national traits to offer her child for adoption. But in other cases, where adoption might prove the easier solution on this basis, one still wonders about the common, almost psychotic, compulsion of some childless women to prove their femininity by bearing their own child even if fathered by a donor's semen. Perhaps there is something innate in woman that makes adoption less fulfilling than a child artificially inseminated. Serious as this question is, Muller urges that we go beyond it to consider the psychology of the normal fertile couple who insist on having their own children by the usual age-old method of coitus. Muller is rather strong in denouncing the "subtle egotistic attachment" of the average married couple who avoid using artificial insemination with a selected donor's semen. He argues that every married couple should consider the distinct advantages of AID and the much greater satisfaction to be gained from knowing that a child is the product of our own deliberate volition rather than of our reflexes. Would we not be happier, he asks, knowing that besides giving this child the very best we can in education and environment we have also given it the best possible set of genes by selecting its genetic father?

Some other interesting considerations can be found in Muller's proposal for a eugenic program of *voluntary choice of the germ plasm* (VCOGP) used for insemination. The present policy in AID is for the physician to select the donor, who remains totally anonymous to the married couple. In fact, both the donor and the inseminated mother remain unknown to each other to avoid emotional complications and any chance of blackmail. The physician's choice is usually pretty haphazard with the local medical students supplying the sperm bank from which he selects on the basis of physical similarities between the donor and the prospective father. (Medical students, following the path of their predecessor Addison Davis Hard, receive from ten to twenty-five dollars for each semen sample contributed to the banks.) Dr. Muller argues that the couple, not the physician, should exercise the choice. With professional counseling, he feels the parents would be better able to choose the genetic endowments of their child, selecting the sperm of "those whose lives had given evidence of outstanding gifts of mind, merits of disposition and character, or physical fitness."

In order for the parents to exercise a choice of the semen to be used, the veil of anonymity that has until now protected the donors would have to be removed. Dr. Glanville Williams, noted professor of British law, has pointed out two prime reasons for the present policy of donor anonymity: "the desire to protect the donor's reputation (think of the repercussions for his family if his adventures in paternity became common gossip!), and to eliminate the risk of the wife transferring her affections to the donor." The second reason has already provided grist for one melodramatic novel, *The Test Tube Father* by Francis Silvin, and many science-fiction stories of husbands who suffer jealousy and wives who have obsessive dreams about the real father of their child. If, somewhat regretting her choice of a married partner, the wife were subconsciously to choose a donor for the traits she finds lacking in her spouse, there is the distinct danger that she might become emotionally attached to him in a way her husband might not like. On the other hand, women might not be concerned about the hereditary background of the men they marry, whether their families have a history of tuberculosis, cancer, hemophilia, muscular dystrophy, or the like, if they knew they could avoid complications and chose the semen of a donor who was free of these defective genes.

With the anonymity gone, we can also extend a classic problem into a new dimension. If a child conceived by AID learned who his genetic father was, might he not be tempted to find with him a true filial relationship? Expanding on a problem often faced by adopted children, this could pose serious threats to the family life of the child. The triangle of affections and emotions that could arise in this situation is very real.

Muller suggested a way to avoid these problems by storing

all semen at least twenty years, during which time the traits of the donor could be evaluated objectively in terms of his later performance. This eliminates the risk Muller himself ran in trying to select ideal donors from current figures—an early list included Lenin. Waiting twenty years would reduce the possibility of choices being influenced by fads, especially if, as Muller urged, semen not be actually used until the donor had died. This last suggestion eliminates pretty much the psychological and emotional complications that could easily arise if the wife knew the donor personally.

One of the prime traits prospective parents look for is that of intellectual ability. Genetically we are still very much in the dark about the heredity of intelligence. Genius parents have given birth to morons and rather ordinary couples have produced geniuses. One has only to note the family backgrounds of the youthful geniuses who occasionally make the newspapers after being admitted to some university at the age of eleven or twelve, finishing a four-year college program in one year, or graduating with a doctorate in astrophysics at seventeen, to realize that intellectual acumen is a very unpredictable trait. Nevertheless, it is probably "easier, by selection, to change the intellectual or other aptitudes of the population than to change the incidence of disabling diseases or sterility," as Professor James F. Crow has noted. In a 1962 symposium on evolution and man's progress, Dr. Crow suggested that

since society owes so much to a small minority of intellectual leaders, a change in the proportion of gifted children would probably confer a much larger benefit on society than would a corresponding increase in the population average. These potential leaders would probably produce enough change in cultural and other environmental influences to be worth considerably more than the contribution of their genotypes to the genetic average. It has been frequently suggested that when artificial insemination is used, because of sterility or genetic disease in the husband, the donors might be selected from men of outstanding intellectual or artistic achievement. Were this widely practiced, I believe that the occasionally highly gifted children, though probably a small proportion of all the children produced, might still be a most important addition to society.

Among other advantages of "ethereal copulation," to resurrect
Dr. Sims's term of the 1850's, is a practical solution for a husband
who does not produce enough sperm in any single emission to
achieve conception. Oligospermia can be circumvented by combining several samples of such a husband's semen, perhaps with
freeze-storage over a period of months until enough is obtained
for proper insemination. It could offer an out to a man facing an
operation for prostate cancer with consequent sterility. In the
summer of 1969 American newspapers reported a case of artificial insemination with frozen sperm in Italy. The husband traveled considerably and never seemed to be home during his wife's
fertile period. The couple resorted to insemination with frozen
sperm and both rejoiced when the wife gave birth to a healthy
baby. (The newspapers reported this as a first, which it may
have been in Italy, though certainly not elsewhere.) This case
suggests an unexpected side effect: it provides the wife with the
perfect opportunity for infidelity. And, even if the husband were
certain his wife had been inseminated with his semen by a reliable doctor, he might later on have some doubts about the
child's paternity.

Physicist Ralph E. Lapp and others have strongly supported
the idea of men placing their semen in lead-shielded vaults to
insure the availability of an undamaged supply in the event of
nuclear warfare, accidental excessive fallout, or even personal
mishaps to individuals.

How do people today accept the idea of artificial insemination and frozen sperm donors? This is a crucial question both
for society in general and for young people, married or single,
who might consider this in the future. Dr. Alan Guttmacher, a
leading figure at the Planned Parenthood Foundation, is reassuring: "Children conceived by artificial insemination often mean
more to families than children conceived in the normal manner.
These children are *wanted*—often desperately wanted." It seems
then that artificial insemination and frozen sperm banking has
already become part of our culture, even if as yet it is not ex-
tensively practiced by the public. Apparently it is being incorporated into our cultural patterns for what it is in fact, an
extension of child adoption, an anticipation by nine or ten

months of the loving adoption of a child by parents who really want a family.

The space age had brought another dimension to the world of artificial insemination. With man now walking among the lunar craters and preparing for a voyage to Mars and beyond, scientists are beginning to wonder about practical ways to populate these new worlds. Obviously it is far too expensive to send grown men on such expeditions where they are of no use and only excess baggage until they land. It might be more practical to send a half-dozen women astronauts with a freezer well stocked with frozen sperm. When the ship lands the women would inseminate themselves and nine months later deliver a batch of new colonists. Further use of the frozen semen could eliminate moral questions of incest, if not of adultery. Should there be any risk of genetic damage to the women on the voyage, frozen eggs protected in small lead chambers could be used for the insemination. Dr. E. S. E. Hafez has already successfully frozen animal embryos for up to twelve days and then revived them by thawing. Perhaps, on longer space voyages where the trip's duration far exceeds the normal human life span, we will store a dozen human fetuses in a small freezer attached to a computer. Some twenty years before the spaceship arrives at Alpha Centauri or some other solar system, the computer would transfer the frozen fetuses to artificial wombs. Nine months later, decantation would be followed by computer-robot nursing and education. One basic question here, however, is whether these twelve beings can rightly be called humans. Their genetic code is definitely human, but their whole psychological, cultural, and reproductive isolation from the human race on earth raises the distinct question of their constituting a whole new species.

Roughly two dozen American and Russian astronauts have ventured beyond the atmospheric shield of our earth. Each of these men and one woman have been exposed to cosmic radiation from which we are protected here on earth. With radiation one of the prime causes of mutations in the genetic code of our cells, should an astronaut and his wife consider frozen sperm banking for their future children? When space travel becomes a regular mode of excursion perhaps a trip to the frozen semen and egg banks will become part of the standard preparations for the trip.

In recent years artificial insemination has found many very promising applications in animal husbandry. It has been used to produce bastards or hybrids between species of animals that ordinarily do not mate. Thus semen from a zebra has been used to inseminate a mare and little zebroids have been born. Cows and the oxlike Zebu of Asia and Africa have been crossed. Cows and bison (buffalos) have likewise been successfully mated this way. On the other hand, attempted crosses between sheep and goats, dogs and foxes, and the European rabbit and the hare have as yet failed to yield any viable offspring. Jumping over the biological class or phylum barrier with artificial insemination is an intriguing if so far unsuccessful project. Scientists have, though, "fertilized" a sea-urchin egg with sperm of the sea lily. Both belong to the phylum Echinodermata but fall in different classes, the *Echinoidea* and *Crinoidea*. Sea-urchin eggs have also been "fertilized" with the sperm of *Mytilus*, a clam in the phylum Mollusca. In both cases the sperm entered the foreign egg, activating it but without participating in the subsequent development. The egg developed parthenogenetically, by virgin birth so to speak, after being activated. Bullfrogs have been crossed with grass or leopard frogs, but again the development grinds to a halt early in the embryonic stages of gastrulation.

There are literally hundreds of animal species in our world today which are on the verge of extinction. Many of them are represented by only a few dozen or hundred scattered about among the zoological parks of Europe, Africa, and the Americas. In captivity many animals refuse to mate or reproduce. They languish in lassitude and with them may die our hopes of preserving some of our rarest animal species. But artificial insemination now offers the scientist a helping hand in the production of new offspring.

Even in the animal world, artificial insemination has its complications, legal and otherwise. The American Kennel Club, for instance, which protects the pedigree of champion dogs, registers pups sired by artificial insemination only if a veterinary certificate attests to the pedigree and source of the semen. In the equine world we have a horse of a different color. Where horse fanciers are concerned, the Jockey Club of England has strictly outlawed artificial insemination of racing or show stock. Japanese breeders, however, are of a different mind and find no objection to artificially inseminating race horses.

Let me raise one final point in terms of the practical application of human artificial insemination, the possibility of selecting not just the intellectual and physical traits of the offspring, but also of predetermining its sex.

Experimental embryologists have for years been aware of differences in the shape of human sperm. In the human male each cell of the body carries an XY pattern of sex chromosomes plus twenty-two pairs of autosomes or nonsex chromosomes. These pairs separate during the formation of sperm with a Y going into an androgenic sperm and an X going to the gynosperm. Androsperms are somewhat lighter and speedier than the X-bearing, female-determining gynosperm. The male-producing androsperms are likewise round-headed while gynosperms have oval-shaped heads.

In a normal ejaculation of semen androsperms outnumber the gynosperms about three to two. Besides these physical differences, the two types of sperm behave differently in an electrical field. These characteristics offer the experimental embryologist a possible foundation for developing an effective technique for separating the two types of sperm. Among the techniques already in the experimental stage are gelatin solutions in which the two types of sperm are allowed to settle to the bottom at different rates because of their different weights. A quicker method to achieve the same results is to spin the sample in a centrifuge. In Sweden, Lindahl used centrifuge-separated sperm to produce eleven consecutive bulls. The Russians have applied a mild electrical current to a suspension of semen in a U-shaped glass tube as a screening process with partial success. The androsperm will head for the negative pole of the electrical field while the gynosperm will make for the positive pole. Dr. Manuel Gordon at Michigan State University has used this approach to alter the usual fifty-fifty sex ratio in rabbit litters, achieving 64 percent male or 71 percent female offspring with the segregated sperm.

How far off is this prospect for applications in human reproduction? That depends on whom you ask. Dr. Paul Ehrlich, of

Stanford University, figures we will be able to select the sex of our offspring in less than twenty years. Somewhere, then, around 1985 young couples may be able to say with certainty "we are having two children, a boy first and then a girl." Dr. Amitai Etzioni, of Columbia University, has been quoted in *Science* to the effect that this technique will be feasible by 1975 *or sooner*. Dr. E. James Leiberman, a National Institutes of Health scientist, has suggested that we develop a special diaphragm which will act as a selective filter, allowing, say, androsperm to pass through but forming a barrier for the gynosperms, or vice versa. Dr. Charles Birch, the renowned geneticist at Sydney University in Australia, goes one step further to predict the development of a pill which will trigger in a man an immune rejection or suppression of one of the two sperm types.

Dr. Sophia Kleegman, a pioneer of some forty years' experience in fertility research, has uncovered one possible way of preselecting the sex of our offspring. In her work at New York University, Dr. Kleegman has timed insemination to coincide with or preceed ovulation. If insemination occurs thirty-six to forty-eight hours before the egg is released from the ovary, there is about an 80 percent chance of its being a girl. If the insemination is timed to coincide with ovulation or very close to it, the probabilities favor the conception of a boy, again by about 4 to 1 odds.

In the spring of 1970 Dr. Landrum B. Shettles coauthored a very important article for popular consumption on the subject of selecting the sex of your next child. The technique proposed in this *Look* magazine article brings together in a very practical synthesis several basic facts about the behavior of androgenic and gynogenic sperm. The guidelines drafted by Shettles offer a simple, safe, inexpensive, do-it-at-home technique which, on the basis of his own clinical experiences, should produce an offspring of the desired sex eight or nine times out of ten. Some of the basic facts Shettles draws on are that androgenic sperm are fast swimmers, though they live only about twenty-four hours; they are more inhibited by the acidic environment of the vagina, while the gynogenic sperm are more affected by the alkaline conditions in the cervix, uterus, and tubes, more favored by an absence of orgasm (which increases alkalinity), and slower

swimmers with a life span two or three times that of the androgenic sperm. Shettles' procedure calls for a minimum of equipment: a seven-dollar glucose fertility test tape kit, a douche kit,
and either a bottle of *white* vinegar or a box of baking soda. Increasing the odds for a girl calls for no intercourse during the
two days prior to ovulation, preceding each coitus immediately
with an acidic douche of two tablespoons of *white* vinegar in a
quart of water, avoidance, if possible of orgasm, shallow penetration, and the use of the face-to-face or "missionary" position for
intercourse. Recommendations for insuring a boy include timing
intercourse as close as possible to ovulation, immediately preceding intercourse with an alkaline douche of two tablespoons of
baking soda in a quart of water, deep penetration at the moment
of male orgasm, and vaginal penetration from the rear. For a
male offspring, orgasm is desirable and intercourse should be
avoided completely from the beginning of the monthly cycle until
the day of ovulation.

Since this is beginning to sound like science fiction—which it is
not—I might as well carry this whole thing one fantastic but possible step further into man's future. Before he retired from Iowa
State University where he had achieved some notable success as
an experimental embryologist, Dr. Emil Witschi suggested that
the next logical step in the predetermination of human sex would
be "a complete separation of the male line into gynosperm-
producing individuals and androsperm-producing individuals."
The human female then would have a real choice between two
distinct breeds of men, one that would give her only female offspring and the other capable of producing only baby boys.

Before you toss Dr. Witschi's idea out the window as completely ridiculous, I should mention that Mother Nature, as usual,
has already anticipated the scientist. Dr. Landrum B. Shettles,
at Columbia-Presbyterian Hospital in New York City, has reported the case of a husband whose semen contained 96 percent
round-headed androsperms. Chances for this gentleman producing a female offspring are only one in twenty-five.

Whatever techniques we finally end up using in the next decade or two to predetermine the sex of our offspring, even the
smallest step in this direction is likely to raise serious complications and consequences in our social realities.

The usual reason offered for predetermining the sex of our offspring is parental preference. This somewhat superficial reason could lead to some serious complications. If parents simply want to balance the sexes in their small families, fine. But what happens if, for a variety of causes or reasons, some parents decide they want children of only one of the sexes. Quite often among today's college students, both young men and women alike, we hear that with the sexual revolution on the college campus and the new freedoms they certainly do not want any girls when they marry—too many problems and worries. Several sociological studies of parental attitudes have indicated the young parents favor male offspring by between 55 and 65 percent over daughters. An imbalance of the sexes in favor of males, according to Columbia University sociologist Amitai Etzioni, would "very likely affect most aspects of social life." Dr. Etzioni, for instance, pointed out that men vote "systematically and significantly more Democratic than women." With the Republican party steadfastly losing support in the past generation, a predominance of males in the voting population could undermine our two-party system. Women are greater consumers of culture, more regular in church attendance, and typically more involved in the moral education of the children so that "a significant and cumulative male surplus will thus produce a society with some of the rougher features of a frontier town," according to Dr. Etzioni. In another likely consequence this sociologist foresees an increase in interracial and interclass tensions because minority groups and the lower socioeconomic classes tend to be more male-oriented than the rest of society. A high surplus of young men in these groups would prompt a search for mates in other classes and groups with the inevitable problems this creates. Other possible consequences Dr. Etzioni foresees, should we shift our sex balance in favor of the male, are some delays in the age of marriage, an increase in prostitution and homosexuality, and a rise in the number of bachelors. To these Albert Rosenfeld, author of *The Second Genesis,* adds the possibility of a switch to polyandry should we end up with too many men. While these dangers are not apocalyptic, they do prompt a question about whether or not the disadvantages are likely to outweigh the advantages in this situation.

Generally speaking there are fewer tensions, problems, and complications when the sexes are nearly balanced in a population. Nature itself has evolved some fascinating strategies to achieve a balance of males and females at mating time. Specific figures on the primary sex ratio, the ratio of boys to girls at fertilization, are not available, but serious estimates indicate that as many as 150 to 160 males are conceived for every 110 females. This is roughly a 3 to 2 ratio, paralleling the 3 to 2 imbalance of the two sperm types. For many genetic and physiological reasons, the male fetus is much more susceptible to miscarriage and malformations. Many zygotes fail to implant in the wall of the uterus a week after conception, and many fetuses, mainly males, abort during the early months of pregnancy. Toward the end of the second month of pregnancy four male embryos miscarry for every one female that aborts. As a result approximately one-third of the males conceived never are born while only about 10 of the 110 female embryos suffer this fate. The secondary sex ratio, at birth, then generally comes closer to a balance. In the United States it is 106 white males and 103 black males for every 100 females born. In Greece 113 males for 100 females and in Cuba 101 to 100. From birth on the ratio drops. Wars, accidents at work, ulcers, heart attacks all seem to take a preferential toll on the male. Statistics from England and Wales for 1960 show a 106 to 100 ratio at birth, 100 to 100 in the thirty age bracket, 90 to 100 about the age of fifty-two and 50 men to 100 women at the age of seventy-five. In the United States there are about 119 females for every 100 males in the sixty-five to sixty-nine age group.

About twenty diseases are known to be directly inherited on the sex-determining chromosomes as sex-linked. These diseases hold promise of possible eugenic control in the context of sex selection. To illustrate this, let me try to compress the essentials of genetics into a few sentences and then move on to a case in hand.

Simple inherited characteristics, such as eye color, hair color, the production of certain proteins which the body uses for growth, and the like are usually controlled by pairs of genes arranged in series. Each pair of genes takes a particular process one step along the assembly line toward the final product. These gene pairs, sets of alleles, are located on the twenty-two paired body chromosomes, autosomes, and on the single pair of sex chromosomes, XX or XY. A person can have two genes for the same kind of character, two genes for brown hair, for instance, or two genes for normal blood hemoglobin. But the two genes may also be different, i.e., one gene may be for brown hair and the other for red hair, in which case the genes might blend their products to give an auburn or reddish brown. The brown gene might also dominate and override the red gene, then known as a recessive, or vice versa. Blending and dominance are but two of the ways paired genes (alleles) can operate, but for us here the dominant/recessive combination is of concern. In order for a person to show a recessive trait physically, he must have two recessive genes for that trait. Blue eyes are recessive, and every blue-eyed person carries two recessive genes for blue eyes in his cells.

Let me add one further point. Twenty-odd diseases are sex-linked and located on a sex chromosome, on the long arm of the X chromosome. The Y chromosome is missing this long arm so that with sex-linked traits, girls receive two alleles, one on each of their paired X chromosomes, while boys receive *only one gene*, the Y carrying no matching allele in this case. Thus a girl can have one normal sex-linked gene and a recessive defective allele for hemophilia and still not show the disease; she is a *carrier*. While a boy with a single recessive gene for hemophilia has nothing to counter it, with the result that, even though the gene is recessive, the boy will have the disease.

Let us apply this now to the question of sex selection. Take the case of a husband and wife where the husband has normal blood and the wife is a carrier for hemophilia—she has one recessive "bad gene" for hemophilia masked by a dominant "good gene" for normal blood. Their male offspring would receive a Y chromosome from the father with no gene for this blood condition, good or bad. The determining gene then comes from the mother, in this situation a fifty-fifty chance of it being normal or recessive hemophilic. Thus on the average one-half of this couple's boys would be hemophilic. The girls would receive a normal X gene from the father and either a normal or hemophilic gene from the mother. All the girls would have normal blood but one-half of them would be carriers of the disease like their mother. If this couple decided to have only girls, the disease would not show up though the defective gene would still be in the gene pool, carried by one-half of the daughters. If these daughters followed through and also had only daughters, the disease again would not appear unless one of them married a hemophiliac.

A variation on this would be even more advantageous, if the husband were a hemophiliac and the wife had two normal genes. All daughters would be carriers and all the sons would be free of the defect. By having only sons, the gene would be eliminated from the gene pool of all their descendants unless reintroduced by marriage.

With the selection of the sex of our offspring we can then hope to eliminate or control, by negative eugenics, twenty-some sex-linked genetic disorders, among them A and B type hemophilias, red-green color blindness, several enzyme deficiencies, and one form of muscular dystrophy. Writing in *The New Scientist*, Drs. Edwards and Gardner commented on the implications of this:

> The elimination of these disorders in one generation, by judicious choice of the sex of the offspring, would not only be of direct benefit to that generation, but would benefit the race for generations to come. More immediately, the ability to determine the sex of domestic animals would be of enormous practical importance to farmers.

And so the biological revolution moves onward. Man begins to create a new technology, to give sex a helping hand, and in so doing we open the door to a whole new world.

As Robert Heilbroner points out in his preface to Seligman's *Most Notorious Victory:* "Technology is altering life to its existential roots before our very eyes." The new technology of human reproduction, far more than any other development, is creating today a whole new culture, if for no other reason because it cuts to the very foundation of human society, the family. Whether it be the end of traditional private capitalism predicted by John Kenneth Galbraith in the emergence of a "technostructure," the alienation brought on by an inability of our youth to meet the demands of the ego in the "technological society" surveyed by psychologist Kenneth Keniston, the instant media culture of Marshall McLuhan, or the technological determinism of Jacques Ellul, one fact is rapidly permeating our consciousness. To paraphrase Victor C. Ferkiss, author of *Technological Man: The Myth and the Reality,* modern technology, especially in the area of human reproduction, is in one way or another *the factor* that is making a radically new civilization and culture possible and necessary. Technology, however, is not merely *the* trigger of a new culture, it is also and perhaps more important "providing its organizing principles."

The reproductive technology we have just surveyed already reveals some of the orientations or trends our new culture will likely take. These trends will, I hope, become much more apparent as we move through the next four chapters. By the time we plunge into our last chapter on human sexuality and its future, I believe they will be more obvious to the reader than I could ever make them with italics, bold-face print, or two-inch headlines.

As a sort of preparation which may help the reader to perceive the overall pattern and not get lost in the forest of minutiae, I would like to express very briefly, in bald blunt statements, some of the key trends or orientations I see in the developments reviewed here. For clarity's sake I will group these trends in three areas: those dealing with the individual, those touching the familial structure, and those involving the abstract world of morality and social constructs, though there is considerable interlocking of these.

I. Among the trends imminent for the individual sexual person, it seems most obvious from what we have already said that in human society sexual intercourse is increasingly moving out of the reproductive sphere. Reproduction can and is being achieved with increasing frequency without the benefit of sexual intercourse. Both these trends are clearly contained in the fact that human reproduction is rapidly becoming more and more a matter of deliberate choice on the part of parents. But this also implies that sexual intercourse must now be more fully integrated into our lives with *some meaning other than the mere reproductive.*

II. In the area of the family, the new technology of reproduction suggests that, while the monogamous couple marriage will likely remain dominant in our culture, it will at the same time be modified in its exclusivity by the appearance and growing acceptance of other forms of male/female relationships and parenthood. Further the stability of the monogamous marriage cannot continue to rest on the rather simplistic base of biological (sexual) fidelity or exclusivity. Today and tomorrow it will have to broaden its base to a more profound and more human interpersonal relationship if the monogamous marriage is to survive in something more than mere name and appearances.

III. Finally, we are rapidly coming into situations where we cannot transplant the moral dictates and categories of generations past into our new culture. The moral categories of adultery, infidelity, incest, fornication, and the like are an *organic* part of a dying pastoral culture which no longer speaks to modern technological man in any kind of intelligible language. The *fundamental* human values of that system *must then be preserved not by mere translation* into a new language but *by a radical transformation*, or better *by an evolutionary growth* out of adolescent or preadolescent conceptions. Legally and socially, we will have to develop new meanings for classic concepts of parent, father, mother, son, etc., often *far beyond* anything we have dreamed of as yet. Sexual morality, which was so clear, neat, cut and dried in the past, focusing as it did almost exclusively on genital sex, will have to be radically thought through by men *and women*, starting always from a deep historical, Judaeo-Christian perspective so that what is of lasting value in past moral applications will be able to mature as man matures in his understanding of his nature.

These are the trends as I see them.
Where might they lead us?
This question will be explored in the final chapter, but first we must build a stronger foundation and learn whether the trends stated here without qualifications will hold up as valid prognosis.

TWO

WOMBS OF
GLASS AND STEEL
How to Decant
Your Baby

〽〽

Scientists call it an "extracorporeal membrane oxygenator," this inexpensive and, of course, disposable contraption about the size and shape of a quart juice can. Through an imposing array of control valves, flow monitors, and a heart pump life-giving oxygen and liquid food are sucked into the silicone rubber bag coiled inside the apparatus. Once there the dissolved oxygen and food move through the semipermeable membrane to the blood flowing on the other side while carbon dioxide diffuses in the opposite direction. Alongside the E.M.O., in a warm moist incubator, a frail bluish human baby hovers near death with hyaline membrane disease, a serious lung disorder. The doctors watch their gauges and instruments. The baby perks up. His color changes to a healthier pink. The proper amounts of oxygen and carbon dioxide flow back and forth through the umbilical arteries and veins between infant and machine. The baby had stopped breathing and this is a last-ditch effort to save its life. For six hours the E.M.O. works away with success. But then the end comes. The inevitable has been postponed a few hours, but not enough to allow the infant to gain strength and overcome its crippling disease.

Sometime later the same doctors watched anxiously as another patient was connected to the machine. This time it was an unborn lamb floating in a tanklike womb that simulated the uterine environment from which it had been removed a few moments before. Once connected to the machine Dr. Theodor Kolobow had invented, the unborn fetus was put through many tests to learn what could be done to improve the E.M.O. Dr. Joseph Pierce, a veterinarian, and Dr. Warren Zapol injected X-ray opaque dyes into the fetal lamb's neck vein and, by X-raying its circulatory system, gathered valuable information for thirteen hours before the lamb died.

The cause of death in both cases was not a failure of the machine, but rather infections resulting from contamination of the system. Even so, the E.M.O. artificial womb has supported life in three dozen lambs for up to two and a half days, the world record as of spring 1969. It has "given birth" to at least one lamb but after nursing it for only four hours.

The work of Dr. Kolobow and his associates at the National Heart Institute, Bethesda, Maryland, is complemented by the research of at least a dozen other scientists around the world. Since Dr. J. A. Thomas first used a modified heart-lung machine to keep calf fetuses alive in France, other scientists have tried to perfect the technique. In Canada, for example, Dr. John C. Callagan has pioneered work with lamb fetuses.

In 1968, at the King's College Hospital, London, doctors tried another tack in the development of an artificial womb. The umbilical cord of an infant born three and a half months premature was passed through a lightly coiled cellulose tube, bathed in fluid containing oxygen and other nutrients. The infant survived only five hours, but in the process another small step was made in the development of a machine that will eventually allow such premature babies to survive.

Scientists at the Stanford University School of Medicine have come about as close as anyone thus far to creating an effective artificial womb. Dr. Robert Goodlin's machine is quite different from that devised by Kolobow. It consists of a thick steel cham-

ber with a small round peephole in which a saline solution can
be saturated with oxygen under about the same pressure a deep-
sea diver experiences at a depth of 450 feet, 200 pounds per
square inch. At this pressure, oxygen is literally forced through
the skin of the fetus. But without a placenta to allow diffusion,
carbon dioxide and other waste products quickly build to toxic
levels and the fetus dies within forty-eight hours. Despite his
very limited success, Goodlin is convinced that all the obstacles
will be overcome and eventually, perhaps in the near future, we
will have available a machine that can save newborn babies suffering with hyaline membrane disease as well as the premature
fetuses which now die because their lungs, at anything under
twenty-four weeks after conception, cannot supply enough oxygen to the body.

While some scientists are struggling to learn more about the
requirements of the fetus during the last few weeks or months
before birth, others are tackling the problems involved in creating an artificial womb that will work with fetuses in their earlier
stages.

At the Strangeways Laboratory in England, Dr. D. A. T. New
has grappled with this problem for several years. His approach
involves removing mice embryos from the womb when they are
about 2 mm. long. Placed on a clot of blood plasma with a drop
of nutrient solution, these fetuses continue to grow. At 2 mm.
they have a rudimentary heart. Their nervous system is a mere
furrow along the back side. In two days they will quadruple in
length, and develop all the main organs: brain, spinal cord, eyes,
ears, intestine, kidneys, limb buds, head, and tail. Their tiny
hearts will begin to beat. But again, as with near-term fetuses,
the lack of a placenta proves fatal. Mice have two placentas.
An early yolk-sac placenta, derived entirely from fetal tissues,
can by diffusion nourish the embryo during these two days and
eliminate its waste products. But then a placenta must develop
involving maternal tissues so that the embryo can draw its nourishment from the mother and get rid of waste products through
her system.

Another pioneer in this field, Dr. Chamberlin, has worked with
human fetuses in both England and Washington, D.C. In his initial experiments, Dr. Chamberlin used eight living human fe-

tuses, weighing between 300 and 980 grams, which had been obtained during hysterotomies for therapeutic abortion. Seven of the fetuses were removed from the womb with the amniotic sac intact while the other was placed in a normal saline solution at the operating table. All the embryos were kept in tanks immersed in artificial amniotic fluid to prevent regular breathing from starting. Within twelve minutes after removal from the womb, the umbilical blood vessels were connected to perfusion equipment, a combination heart-lung-kidney machine. The largest fetus, a male from a fourteen-year-old girl, survived the longest in this series of experiments. As the *Ob-Gyn Observer* reported: "A brisk spontaneous flow [of blood] was noted 22 minutes post partum; the fetus was kept on the circuit for 5 hours and 8 minutes. . . . Only when a cannula slipped out by accident and could not be reintroduced was the experiment halted."

The high hurdle along the road to an effective artificial womb is that marvelous and mysterious combination of maternal and fetal tissues known as the placenta. It acts as a selective filter passing certain substances, amino acids, vitamins, oxygen, and sugars from mother to the fetus, but acting as a filter to screen out many proteins, blood cells, and other substances which might trigger an immune rejection in the embryo or cause it harm. It took scientists five years of intensive experiments to understand the physiological processes involved in the simple diffusion of oxygen and carbon dioxide across the placental barrier. Much of this work was done at the Department of Experimental Medicine at Cambridge with pig embryos and a variety of artificial placentas. To understand how liquids and solids migrate selectively across the placenta will likely take scientists another ten years.

In solving the problem of an artificial placenta we can use a variety of techniques. One of these presumes on that rather peculiar critter, the opossum. A marsupial, the opossum gives birth to very immature young some twelve to thirteen days after conception. At this stage, a litter of ten or twelve can fit comfortably into the bowl of a tablespoon! Thrust into this cold world, the minute infants must drag themselves from the birth canal up to their mother's pouch, tugging on her body hairs with their strong forearms as if they were working their way through a dense forest. Once safely in the maternal pouch, the infants attach themselves to a nipple and nurse for the next seven or eight weeks before moving out on their own. This life habit offers the scientist a unique opportunity to study fetal development outside the womb. In a plastic robot pouch, engineered by scientists at the Marquardt Corporation in Van Nuys, California, twelve infant opossums of various ages can be studied as they mature in their see-through compartments. As each fetus nurses hungrily on his nipple, scientists can feed him selected foods and drugs, observing his reactions and the effects of the food or drug in a direct way which is as yet impossible with mammals.

This is a useful technique, but it answers only a few of our questions about prenatal life and none at all about the actual functioning of the mammalian placenta. Those answers must come from experimental work done directly with actual mammalian placentas. Thus, in Paris, Professor Maurice Panigel has studied perfusion of natural but isolated placentae. Other scientists have joined in exploring the physiological mechanisms behind the placenta. To learn more about the placental-uterine world which shelters and nourishes the fetus, Dr. John Mc-Cracken, at the Worcester Foundation for Experimental Biology in Massachusetts, has brought the uterus out into the open. Instead of trying to study its functions when isolated from the animal altogether, as Panigel has done, McCracken has grafted the uterus and ovaries of a ewe to its neck. In a future experiment he hopes to implant a sheep zygote in the relocated uterus and observe its development there.

The technical difficulties of duplicating the mammalian placenta are considerable, primarily because practically all artificial substances, even those coated with slippery silicone or teflon, easily damage blood cells. Another problem is the difficult task of connecting plastic tubes to the fragile, tiny veins and arteries of the umbilical cord that leads to the placenta from the fetus.

Dr. Kermit Krantz has devised an intriguing apparatus at the University of Kansas Medical Center which sandwiches a living human placenta, with its partially amputated umbilical cord still attached, between two clear plastic disks. Tubes simulating the blood supply of the fetus are then connected to the umbilical arteries and veins while a near hundred tubes enter and leave the bottom of the sandwich with blood to simulate the maternal circulation. With this ingenious device, Dr. Krantz can study the actual functioning of a human placenta as food, oxygen, hormones, and antibodies pass from the mother's side to that of the fetus and as waste products leave the fetal circulation. Malfunctions of the placenta during pregnancy can easily reduce a potential genius to a mentally slow individual by allowing toxic wastes to build up in the blood. Krantz hopes that by learning how the placenta can malfunction we may be able eventually to prevent such disorders and perhaps even improve the chances of a potentially mediocre embryo escaping this fate by improving the functioning of its placenta.

A 1969 editorial in *The New Scientist* was strongly optimistic in its appraisal of the present state of the art, suggesting that "the development of the 'perfect' artificial placenta can only be a matter of time." When the "perfect" artificial womb is developed it will probably not look like any of the equipment thus far in use, though it will very likely incorporate certain elements from Goodlin's hyperbaric chamber, Krantz's living placenta, and Kolobow's E.M.O., refined and complemented by as yet undeveloped components.

The issue of raising a human fetus in an artificial womb, thereby relieving Eve of the "animalistic burden of pregnancy," has been joked about for many years. Yet Aldous Huxley was some five hundred years off in his forecast of the brave new world. In the heart of the Depression years he tried to picture the world of *six hundred hence*, a world that is *already with us*. The satire came off beautifully, for Huxley was a keen observer of his own civilization. In pointing out to us where certain trends *already evident* in our culture of the thirties might lead, Mr. Huxley undoubtedly drew some insights from his brother Julian, no mean biologist and philosopher. The opening pages of *Brave New World* reflect Julian's interest in experimental embryology which in the early thirties was a very embryonic science.

Mr. Huxley introduced his readers to his brave new world by recalling a day in the year 632 A.F. (After Ford), when the Director of the Central Hatchery and Conditioning Center guided a group of new students on a tour of the assembly line where human beings were being mass-produced. The entire nine months of development was carefully supervised by trained technicians in a way that completely bypassed sexual intercourse and personal parenthood. Eggs and sperm were collected in a sterile, scientific procedure, fertilized, and placed in large bottles with a solution specific for the type of individual desired. Decantation Day would come for these "babies," only to be followed by further psychological conditioning until at length each child was properly prepared to enter its predestined class in society.

Devotees of the best-seller list joked about Huxley's satire, much as Frenchmen a hundred years earlier had laughed at Diderot when he recorded *The Dream of d'Alembert.* "A warm room with the floor covered with little pots, and on each of these pots a label: soldiers, magistrates, philosophers, poets, potted prostitutes, potted kings. . . ." seemed to Frenchmen just a bit too fantastic to be possible or realistic as a prophecy. Yet has the situation changed much in a century? Many Americans today react with the same smiling benign incredulity—a 1969 Harris survey of Americans revealed that only 3 percent of the population had even heard of the rather common practice of artificial insemination!

Perhaps this is why, despite the warnings of Diderot and Huxley, the public reacted with shock, horror, moralistic tirades, as well as the threat of criminal suits when in 1959 Dr. Daniele Petrucci calmly announced that after more than forty failures he had finally fertilized a human egg *in vitro* and kept the resulting embryo alive in an artificial environment for twenty-nine days. The new and unexpected invariably brings an emotional response, especially when the new and unexpected threatens the man in the street in a very personal way. The forty-three-year-old father of two children terminated his experiment at twenty-nine days because, in his words, the embryo had become "deformed and enlarged—a monstrosity."

Official reactions echoed the public indignation of the newspapers, speaking of a Frankenstein monster and "The Thing." *L'Osservatore Romano,* the Vatican's semiofficial daily, called upon Petrucci to cease and desist. Petrucci himself was very concerned about the implications of his experiment. His prime motive was to find a way to culture organs that would not be subject to the rejection phenomenon when transplanted. Conferences with

Vatican officials made him wonder whether he might not have committed a double-edged sin, creating human life in an unnatural way and then destroying it, but Petrucci was not convinced that his experiments were all that immoral. For a few months he continued, keeping one experimental baby, a female, alive for a full fifty-nine days before it died owing to a technical mistake.

In November of 1961 the Soviet Union invited Dr. Petrucci to visit Moscow. After two months of conferences, meetings, lectures, and lab demonstrations with Soviet scientists at the Institute of Experimental Biology, Petrucci returned to Bologna, Soviet medal in hand, fully aware that a Russian research team would pursue his experiments. A year passed and the experiments at the Institute did not work out. The Russians invited Petrucci back to Moscow for another two-month session.

The work in Moscow was carried out by Dr. Pyotr Anokhin, a brilliant septuagenarian from Leningrad, with Dr. Ivan Nikolaivitch Maiscki, director of the prestigious Institute of Experimental Biology at the Academy of Medical Sciences, supervising. The six-story Institute at No. 8 Baltiskaya Street in the northern suburbs of Moscow is not an imposing or very modern building by any standard, but the scientists working there have a very fine reputation. In early 1966 they announced that they had managed to keep over 250 human embryos alive beyond Petrucci's record of fifty-nine days. One of these fetuses was reported to have lived six months and to have reached one pound and two ounces before dying.

Unbelievable? Perhaps. . . . Yet Russian scientists have pioneered many advances in the area of reproduction. They have repeatedly hinted at remarkable successes in research with artificial wombs, and Professor Anokhin has been working in this area for many years. Our skepticism might be modified if we recalled the equally incredible success of Demikhov and his coworkers in transplanting dog heads. Still, very little in the way of scientific detail has been forthcoming and one cannot help but wonder what the actual state of the art is today in Russia. The 1966 report suggested that the Russians hoped in the near future to "give birth" to the first human ever to spend his entire prenatal life in a laboratory womb. But as of late 1970 this apparently has not come about.

There is likewise considerable skepticism within the scientific community regarding the work of Dr. Petrucci. His claim is not impossible if one compares it with Dr. New's work at the Strangeways Laboratory in England. As noted above, New has kept mouse embryos alive for two days without benefit of a maternal placenta. This is about one-tenth the normal mouse gestation period of twenty-one days. Twenty-nine days is likewise about one-tenth of the human gestation period of 280 days. The skepticism, however, comes from the fact that Petrucci has not let anyone see the details of his results, and this is always grounds for suspicion in the scientific community. In the early sixties Petrucci claimed to have accomplished what other scientists around the world have been struggling to do for a decade and more, a point of criticism raised rather bluntly by Dr. R. G. Edwards, a British pioneer in this area.

Fifty years after Karl von Baer described the dog's egg in 1827, scientists began groping for a way to fertilize the mammalian egg *in vitro*, as they say, though the Latin for "in glass" indicates here the use of an artificial environment rather than a glass test tube as many think.

Spallanzani had success fertilizing frog eggs outside the female body, but then frogs do it that way naturally. Both the frog egg and sperm are adapted to achieving fertilization outside the female's body in ordinary, often polluted pond water, so the trick was not especially difficult despite its pioneering character. Many high school biologists today can fertilize frog eggs in their labs. But fertilizing an egg from a mammal or human is another game altogether. The mammalian egg has been carefully adapted over many centuries of evolution and natural selection to a very delicately balanced environment in the ovarian follicle, the fallopian tube, and the womb. Likewise, the mammalian and human sperm has been adapted through millennia by natural selection to survive in the environment of the vaginal canal and the uterus and achieve fertilization in the fallopian tubes. So finely adjusted are these germ cells to their milieu that even the slightest deviation can prevent the sperm from fertilizing the egg. On its way to an egg the human sperm must travel the equivalent of a trout swimming five hundred miles upstream. Along the way it faces many obstacles and only a few of the three hundred million sperm that start actually reach the egg. A slight rise in the acidic condition in the vaginal canal or the uterus, for instance, can wipe out millions of sperm in minutes and even eliminate the possibility of fertilization. The scientist must duplicate almost exactly this delicately balanced internal environment of the female reproductive tract, or fail in his attempt to achieve *in vitro* fertilization.

Halfway between the time Diderot recorded d'Alembert's dream and Huxley toured the "brave new world," in 1878 to be exact, Schenk placed a few rabbit eggs he had removed from their follicles in a culture dish. As the eggs floated in follicular fluid along with a small piece of uterine mucosa, Schenk dropped on them some sperm he had drawn from the epididymal tubes above the testes. Though he did not achieve actual fertilization, he was able to note at least one of the early stages in that process, the shedding of the granular layer of nurse cells surrounding the egg as it leaves the ovarian follicle. This layer must break up before the sperm can reach the egg. Crude as it was in imitating the internal milieu, Schenk's experiment did inspire some further investigations in the years that followed.

Between 1878 and the 1960's nearly two dozen scientists in a dozen different countries pursued the elusive goal. Among them were Onanoff in 1893 with rabbits and guinea pigs, the incomparable F. R. Lillie working at the Woods Hole Marine Biological Station in the twenties and thirties, Yamane and Gregory Pincus with rabbits and humans in the thirties, Krasovskaja with rats and rabbits in 1935, John Rock and Menkin with human eggs in the forties, Moricard with rabbits and humans in the early fifties, and Austin, Yanagimachi, Chang, Dauzier, and Thibault with golden hamsters and other mammals in the fifties.

In 1930 a man who later gained worldwide fame as one of the pioneers in developing the contraceptive pills claimed to have actually achieved the elusive goal of a test-tube conception with a mammalian egg. Though Gregory Pincus was very cautious in his official report he did indicate his belief that some of the tubal rabbit eggs he had incubated in a test tube with sperm had been fertilized. In 1934 Pincus and a fellow scientist, Enzmann, transplanted some rabbit eggs to another female after they had been carefully washed free of sperm (rabbits ovulate only after mating) and been reexposed to other sperm *in vitro*. When offspring were born, Pincus argued that fertilization had occurred in an artificial environment. In both cases and with some hindsight, later researchers have argued that most likely the goal of a true test-tube conception eluded the brilliant Pincus.

Dr. John Rock, author of *The Time Has Come*, pioneered further work with human eggs in the 1940's. Working at the Boston Hospital for Women with Miriam Menkin, John Rock cultured human ovarian eggs *in vitro* for twenty-four hours to induce them to mature. (The human egg is ovulated while it is still not fully mature; complete maturation, expulsion of the second polar body, occurs only after the egg is either fertilized or activated.) The mature eggs were then cultured for another forty-five hours in the presence of sperm. Altogether four of the 138 eggs cleaved into two and then three cells. Whether these eggs were actually fertilized or not is still a question today, mainly because scientists are conscious of the fact that eggs cultured *in vitro* frequently undergo a spontaneous parthenogenic development. In a type of virgin birth eggs can begin to develop into adults without being fertilized by sperm.

Landrum B. Shettles, an obstetrician at Columbia-Presbyterian's Sloane Hospital for Women in New York City, tried a different approach in the 1950's. Following the suggestion of Schenk, Shettles incubated human ovarian eggs removed during operations with semen and pieces of mucosa from the tubes. His series of phase photographs showed the development of an egg into a solid mass of cells, a morula. But even the best of these pictures did not rule out the possibility that the eggs were artificially activated by the culture technique and developed parthenogenetically without the aid of any sperm.

At this point in the search, a major discovery was made almost simultaneously, in 1951, by Colin R. Austin of Cambridge University and M. C. Chang working at the Worcester Foundation for Experimental Biology in Massachusetts. These two scientists found that rabbit sperm had to be triggered or "capacitated" by exposure to fluid from the reproductive tract of the doe before they could penetrate the egg. (Capacitation is still a very poorly understood physiological change, but some facts are known. Without capacitation the sperm of frogs, rats, hamsters, ferrets, sheep, and pigs cannot penetrate the layer of nurse cells and the *zona pellucida* which enclose the egg.)

The pace quickened considerably after 1951 with success seeming to dangle in the air waiting to be plucked.

In 1953 Venge reported fertilization of rabbit eggs and, after transfer to recipient foster mothers, the birth of several healthy litters. Despite this seemingly conclusive proof of fertilization, Venge concluded that the development in his experiment might have been due to chance, parthenogenic, and not necessarily the result of actual fertilization of the eggs. In 1959 Chang repeated Venge's work with some variations, and for the first time in many experiments the evidence for *in vitro* fertilization for a mammal was incontestable.

At least two other research teams were able to repeat Chang's experiment with the rabbit. By 1968 the golden hamster had also produced offspring conceived in a culture dish and implanted in a foster mother. In December of 1968, David Whittingham of Sydney, Australia, reported the "first successful attempt to get sperm to penetrate and fertilize mouse eggs outside the female's body." Transplanted to a foster mother, the eggs produced nine normal fetuses. In the same year Dr. Wesley Whitten of the Jackson Laboratories at Bar Harbor, Maine, and Dr. John Biggers of The Johns Hopkins University carried mice from the zygote stage, the fertilized egg, to blastocyst, the point of development during which the embryo usually implants itself in the wall of the uterus. For the first time, man observed all the stages of development up to implantation in a single mammal, and the crucial fact was that this was done with a culture medium whose every ingredient was known and measured.

All of this made the scientists quite optimistic that they were on the verge of a breakthrough in achieving *in vitro* fertilization with human eggs. Cambridge University soon became the focal point for this research. As early as 1960 Dr. R. G. Edwards had begun to mold a research team of scientists interested in making the breakthrough once and for all a certainty no one could dispute. They began carefully and slowly, first plotting out and timing the sequence of events from the resting stage the egg is in when it breaks forth from the ovarian follicle to the moment it matures, releases the first polar body, and becomes fertilizable.

This they did by culturing for thirty-six to forty-three hours eggs gathered from women in various stages of the menstrual cycle. They now knew what was needed to bring the human egg along in its development after ovulation or collection to the point where a sperm could break through its barriers and achieve fertilization.

By 1968 when complications with the egg had been mainly resolved, scientists turned their attention to the sperm. What is the trigger, the key that capacitates a human sperm so that it can fertilize an egg? Edwards and his co-workers tried a number of approaches. They washed human sperm, aged them with uterine tissues and bits of fallopian tubes, and then tried to fertilize some eggs. Of the ninety eggs they used at this time only seven showed any sign of being fertilized. And with these the old possibility of simple egg activation and parthenogenic development without benefit of the sperm could not be ruled out.

Hovering just beneath the surface of this brief outline of medical research into test-tube fertilization is a very basic question with both philosophical and biological ramifications. Teilhard de Chardin expressed it very well and repeatedly when he noted that "nothing, absolutely nothing enters our world except by way of a *birth*." Life is a *process* and if we are to appreciate and understand our world today the dimension of *process* must become a natural and so to speak habitual springboard for everything we try to picture or conceptualize in our world. In his master tome *Animal Species and Evolution,* the biologist and philosopher Ernst Mayr has highlighted the fact that no greater revolution has ever impacted on the biological sciences than the shift from a fixed cosmology in which we view everything as essentially unchanged and *possessed of a fixed complete nature* throughout its existence to what Teilhard so aptly termed a cosmogenesis, a world vision in which all things including man's nature are in *process*.

The *process* of fertilization by which a new living organism comes into existence is a perfect example of this process reality. People unacquainted with the biological sciences, and, sad to say, many students of biology, forget or never quite become habitually conscious of the fact that conception is *not* the instantaneous union of male and female germ cells. Conception is *not* that moment in time when a sperm enters the egg and fertilization is accomplished.

This gross oversimplification is actually the root of many, if not all, the controversies and arguments about *in vitro* fertilization up to the time Edwards and his co-workers sat down to document the *process* involved in the maturation of the human egg ten years ago. For many biologists echoed the common sentiment that conception is, plain and simple, the entry of the sperm into the egg. If a second criterion was to be added it would be the division of the egg cell into two daughter cells.

The mammalian egg as it bursts from the ovarian follicle has not yet completed its basic development. Each germ cell undergoes two cell divisions, known as meiosis, during which the chromosome number is reduced to half. The first division occurs while the egg cell is still nestled among its nurse cells in the ovarian follicle. This division sends one chromosome from each of the cell's twenty-three pairs to each of the two daughter cells. One daughter cell will continue developing into a mature ovum; the other, with its complete set of chromosomes, one from each of the twenty-three pairs, forms a small primary polar body which soon disintegrates, perhaps after one further abortive division. In mammals, the primary oocyte then goes into a resting period waiting for the rupture of the follicle to begin its hopeful journey.

Ovulation is an explosive affair as the pressure of the liquid inside the follicle gradually builds to a breakpoint and the wall of the ovary weakens with the follicle pressing against it. The resting egg bursts from the ovary, surrounded by several thousand nurse cells torn from the follicle wall with the explosion. For a short while the egg floats in the fluids of the body cavity until currents produced by waving fleshy fingers at the opening of the fallopian tube entice it inside. Once inside the fallopian tube serial muscle contractions and tiny hairs, cilia, waving with the symmetry of a chorus of ballet swimmers, push the egg along, stripping it gently of most of its protective nurse cells. In the next twenty-four hours this human egg, about the size of a period printed on this page, must be fertilized or it will degenerate.

Of the more than three hundred million sperm released by the male, only about fifty will finally cluster around the egg as it moves down the fallopian tube. Of these, perhaps one will achieve fertilization. Once the sperm makes contact with the egg, clumping occurs and an enzyme for dissolving protein is released, probably from the tip of the sperm's head. This enzyme helps the sperm worm its way between the remaining layer or two of nurse cells and the *zona pellucida,* a tough outer membrane secreted by the egg and its nurse cells while still in the follicle. Finally as the sperm head comes into actual contact with the egg membrane, or vitellus, the cytoplasm of the egg seems to engulf it almost like a clear cone reaching out in embrace. Within seconds the cell materials of egg and sperm are united.

Though there are many differences in the details of the fertilization process from one animal species to another, the pattern observed in the sea urchin offers a basic overall picture.

Entry of the sperm triggers a chain of reactions which together constitute activation of the egg. One of the very first of these reactions occurs on the surface of the egg. In the sea urchin tiny granules just under the egg membrane begin to break up and explode, starting at the point of contact and spreading out in a concentric wave around the whole egg. Color on the egg surface changes as material from the granules invades the cell membrane, swelling like a sponge absorbing water to raise a new protective covering. This fertilization membrane prevents other sperm from entering the egg. The wavelike explosion of the cortical granules takes between ten and twenty seconds while the elevation of the fertilization membrane is completed within one to three minutes after the sperm touches home.

At the same time other essential changes are going on inside the egg. The pattern of light reflected from the egg changes, viscosity of the egg material increases as does its permeability to water and potassium ions. The electrical potential across the cell membrane drops about ten millivolts for some twenty seconds after fertilization. Other changes have been noted, but these will give you a clear enough idea of the complicated *process* under way with fertilization.

Even as the sperm is activating the egg, the egg nucleus is uncoiling its chromosomes which were duplicated during the resting phase. These divide equally, one batch to form the egg pronucleus and the other to form a discarded second polar body. Both the egg nucleus proper and the sperm nucleus absorb water, expanding like two balloons drifting in a sea of cytoplasm to meet each other. Finally the two pronuclei, each carrying half the original chromosome number, fuse, duplicate their chromosomes, and divide almost immediately into two daughter nuclei. Each daughter nucleus then contains a complete set of hereditary information, half contributed by the father's sperm nucleus and the other half coming from the mother. As soon as nuclear division is complete, the large mass of cytoplasm that makes up the fertilized egg begins to divide so that each of the two nuclei can have all to itself a cell to control. Roughly forty minutes after the sea-urchin egg is fertilized, it shows signs of dividing into two cells. In the frog this first cleavage takes about two and a half hours.

Fertilization is, if nothing else, a perfect example of a biological *process*. It brings home very forcefully the fact that we must learn to think of conception not as some miraculous instant, a split second in which a new creature, a human being, comes into existence. Fertilization as a process initiates a much larger process, that by which each one of us *becomes human during our lifetime*. Ashley Montagu expressed it very well when he said that a human baby is not born with a human nature, but rather with *the capacity to become human*.

Past generations of biologists have thought of fertilization, at times subconsciously, as some magical instant. John Rock, Gregory Pincus, and others set as their sole criterion of fertilization the entry of the sperm into the egg. This obviously is far too simplistic. A sperm, or several sperm, may very well enter an egg. But the egg may be beyond the time where it can respond in a normal way. The process can, and often does, abort.

Even the cleavage of an egg into two or more cells is not a definitive indication that fertilization has occurred. Some eggs, the frog, sea urchin, and others, for instance, can easily be activated by chemicals, mechanical pricking with a glass needle, and similar artificial stimuli. Even without any external stimuli, a degenerating egg may undergo several abortive parthenogenic divisions producing a mass of cells that soon dies.

There is then, despite the common use of the phrase, absolutely no such thing as "the moment of conception." Conception, fertilization as a *process*, takes several hours. Obvious as this is once one is aware of some of the embryological observations, certain implications may not be so evident.

These implications tie in with our earlier remarks on various Greek and Renaissance theories about just when an embryo becomes truly endowed with a human nature. The answer to this question was directly linked in the minds of scholars of those days with the question of when a human embryo takes on a recognizable human form. Philosophically speaking there seemed to be no logic in assuming that a human life principle or soul was present in the embryo until it had the shape and functioning of a human being. Thus biological observations served as the basis for philosophical conclusions. Today some of these early scientific observations and the conclusions drawn from them may appear naïve and unimportant, but they do shed some helpful light on our present attempts to answer these perennial questions with more depth and understanding. Looking ahead a few decades, it is not hard to foresee the scientist and philosopher of the year 2000 searching our own relatively naïve and primitive understanding of human embryology for clues that will help them go even further into the question of what human nature is and when an organism can be considered truly human.

Even as far back as the fifth century before Christ we can find some helpful clues amid the many now dated and erroneous observations and conclusions. In that era Hippocrates suggested that humans are conceived through a union of both male and female semen, though the latter played a minor role in the formation of a human being. In a work attributed to the Father of Medicine, *On the Nature of Children,* the author reports one fetus he studied was breathing only six days after the preformed semen was activated by intercourse. In another work, *On Warm Blooded Animals,* Hippocrates suggests through his disciples that in seven days the human body has all that it needs in the way of major organs and structures. A male embryo though, he suggests, requires at most thirty days for its complete organization with the female embryo being two weeks slower. This very early appearance of the human form in an embryo fitted beautifully Hippocrates' theory that the human is actually preformed in the male semen and its structure only unfolds and expands like a bud coming into full bloom. Our records of Hippocrates' ideas, however, are very fragmentary and we have little to indicate what his thoughts were on when an embryo becomes endowed with a human nature.

Fortunately, our record of another Greek philosopher-scientist is far more complete and helpful on this point. In the fourth century B.C. Aristotle devoted much time to a discussion of human conception, especially in his books *On the Reproduction of Animals* and *On the Parts of Animals*. Aristotle disagreed with Hippocrates about blood being the main component in the semen and with the implication that the semen is fully preformed with a life of its own. For Aristotle neither the male nor the female semen possessed a life of its own, nor did they achieve fullness of life when first united. Rather through the motion of the sexual union the male semen triggers the female seed; the product of the union begins to move, to grow, and to differentiate. It becomes, according to Aristotle, a vegetative form of life, an unformed but differentiating living mass. During this time the primitive organism is animated by a vegetative life principle. As development and differentiation progress, the organism becomes sensitive. An animal life principle then animates its functions. Finally, the organism takes on a truly human form and is then animated by a human life principle, a spiritual soul.

When does this human soul actually enter the body? On that point, Aristotle thought he could give a pretty good answer. First of all, one had to realize that the human life principle cannot be present until the form of the organism requires its activating functions. In plainer words, an organism which is obviously carrying on vegetative or mere sentient animal life functions has no need for a rational soul. Secondly, the human miscarriages Aristotle studied shed some light on the question. It seemed from his observations that certain embryos took on the distinct external characteristics of the male about forty days after coitus while female embryos could not be distinguished until some eighty or ninety days after conception.

If the human soul is not present until it is required by the form and functions of the organism, then the male fetus receives its soul somewhere around the middle of the second month after conception and the female child about the end of the third month of prenatal life.

Aristotle was quite an accomplished biologist for his day and a good philosopher. He undoubtedly was a much better philosopher than many later philosophers simply because he based his philosophical analysis and insights on actual experimental observations of nature. Aristotle was no armchair philosopher, sitting in his library speculating on the nature of the world and of man outside without ever laying a critical eye to that world. Considering his quite logical principle that animating forms (life principles) are related to the functions of the organism they activate plus his actual observations of aborted human fetuses, his conclusions about the male receiving its soul about the fortieth day and the female around the eightieth day are not at all ridiculous. In some respects, Aristotle's view of human development is quite congenial with modern theories of embryonic development, much more congenial than the preformation theory of Hippocrates.

In the second century of the Christian Era, the most celebrated of ancient medical writers, Galen, tried to revive the preformation theory. "It is clear that, together with the seed implanted in the womb, there is a soul, placed there by the Creator, with power to govern the body." This view, in a way, anticipated the theory popularized some years later by Tertullian, an interesting heretic, who believed that the human soul preexists the body and is transmitted to the new offspring in the seed of the parents. A rather facile explanation of how the original sin was transmitted from Adam to his children, this theory, known as Traducianism, was soon condemned by Pope Anastasius II as heretical.

At first, Augustine was attracted by the Traducianist theory, but as he delved into the problem further he found himself very much in agreement with Aristotle. However, in his third-century *Enchiridion* he cautiously points out that he really is not sure when a human person actually comes into being in the womb. Despite this caution and long after his death Augustine's eminence and authority were used as a fulcrum to maneuver Western man into an almost universal belief that the soul is infused by the creator into the developing embryo only when the organism is ready to function as a human body. This view was cogently argued in two apocryphal works attributed to Augustine which gained wide circulation and considerable influence in the Middle

Ages, *A Book on the Spirit and the Soul* and *Regarding Church Dogmas.*

The all-pervasive influence of Augustine, coupled with the revival of Aristotle's writings, led to the firm establishment of the Mediate Animation Theory throughout the Christian world. This theory states simply that the human soul is *not given* by the Creator *at the moment of conception, immediately,* but rather only after a series of developmental stages, sometime after conception when the developing organism has passed through vegetative and sentient or animal phases of life. Mediate animation was so clearly supported by biological observations and philosophical logic that, at the time of the Reformation, the Catechism of the Council of Trent treated it as taken for granted by all, Protestants and Catholics alike, and absolutely certain.

Toward the close of the Middle Ages, when scholastic philosophy was on the decline and improved methods of observation like the microscope replaced crude magnifying lenses, controversy returned to the question of man's origins.

Gassendi, a noted physician of the early seventeenth century, reported his observation of five aborted fetuses, one of which he claimed was only twelve days old and already properly formed and organized, head to toe. If true, this observation would require a revision of the Mediate Animation Theory of Aristotle, Augustine, and Aquinas.

Other evidence, or at least indications, started to pile up. A professor at Louvain University, Thomas Fienus, reported a three-day-old embryo that had a fully formed human body. A Roman doctor by the name of Zacchias argued in 1661 that the soul is present from conception because Catholics celebrate the feast of the Immaculate Conception of Our Lady on December 8, nine months before the feast of her Nativity on September 8, and the Conception of John the Baptist on September 23, again nine months before his nativity. When de Graaf discovered the "egg" (follicle) in 1671 and the sperm came beneath the microscope in 1677, the fiery debate began in earnest. A certain Andryus suggested that some human sperm were actually preformed humans and developed into such while other human sperm were merely seed for insects or worms. Another investigator, Delempatius, said he had seen a tiny human curled up inside the sperm even though his contemporaries laughed at the suggestion. Riolanus, a contemporary of Harvey, reported his findings of three aborted fetuses, one only a month old and fully human in form. In 1724 Bianchi, a celebrated professor at Turin University, gave detailed descriptions of several aborted fetuses in his museum. One of these, fully human in body, came from a certain lady only three or four days after impregnation, while another, only seven days old, had come from a certain most chaste and illustrious Turin matron, one week after her nuptials. This evidence, when coupled with de Graaf's discovery, was sufficient in Bianchi's mind to overthrow once and for all that pernicious Mediate Animation Theory and prove that the human soul is present from the moment of conception. If the human being is preformed in either the egg or sperm, then the human soul must also be there as soon as it is activated.

Besides, with new theories of evolution coming from those atheistic German philosophers of nature it would be better to eliminate any admission that the individual person undergoes any sort of evolutionary development. The philosophers had dared suggest that if the individual evolves, perhaps also the universe is evolving. Preformation of the individual certainly knocked the pins out from under that canard. Or so it seemed.

Still there were those who continued to favor Mediate Animation. As late as 1758 the celebrated author of *Sacred Embryology*, Caniamila, argued that sperm were not even the source of human generation, let alone preformed human beings. "Worms" similar to sperm had been found in the blood and in female lymph and elsewhere. No one had ever found sperm inside an egg. And besides, if the sperm contained a preformed human, "marriage would be obligatory and a vow of virginity would be sinful, inasmuch as it would prevent multitudes of human beings from ever seeing the light of day; polygamy would be imposed by the natural law as necessary, in order to bring all these human beings in the male seed to fruition." And on the argument ran.

Despite the logic of arguments to the contrary, a magnificent array of scientific observation seemed to be piling up in favor of the belief that the human soul is present from the first "moment of conception."

Then quietly the supposed preformed human within the egg

and sperm dissolved beneath the eye of improved microscopes. A glorious empirical construct collapsed. The biologists returned, after some argument, to their original view of a developmental process in which the human body gradually comes into being from an unformed mass of cells containing only certain hereditary information as yet in the main unexpressed in the cells. Philosophers and theologians, on the other hand, who had flocked to embrace the Immediate Animation Theory because the soul must be present from the first moment of conception when that preformed seminal pattern is activated suddenly found themselves without experimental support. Rather than admit their mistake, which was not all that ignominious considering the scientific knowledge and theory of the day, they decided to ignore the new evidence. The result has been a sad one, for our common philosophical understanding of man's origins on which we often base our decisions in medical ethics regarding abortion has been totally divorced from the modern scientific understanding of embryonic development. Our philosophical view of man's origin has in fact ended up in a position diametrically opposite that held by our scientific view of man's conception. This resistance of philosophers and theologians to the overwhelming evidence of biology and the impossible situation it places us in when dealing with questions of medical ethics has been labeled by Canon Dorlodot, a pioneer in reconciling process evolutionary thought with Christian theology in the early part of this century, "one of the most shameful things in the history of thought."

The philosophical question of just where in human development we should place that dividing line between prehuman and human on which we base the new individual's inalienable right to life and the pursuit of happiness has found a very practical application in the experimental laboratory of Drs. R. G. Edwards, Barry Bavister, and Patrick Steptoe at Cambridge. By early 1969 the problem of sperm capacitation had been solved, the development of the egg outlined in detail, and some major steps taken in imitating the environment the egg finds itself in during natural fertilization. At Tulane University in Louisiana, Dr. C. R. Austin and his colleagues had discovered that fluid from the ovarian follicle must be present in the medium if the egg is to mature outside the female body. At the University of

Pennsylvania, Dr. Ralph Brinster outlined other essentials in the ideal medium for egg fertilization: a solution with a salt composition very similar to that of blood plasma with added proteins, glucose (sugar), and pyruvate at specific concentrations. Placing the zygotes in a drop of this medium under liquid paraffin in an atmosphere of 5 percent carbon dioxide also helped.

In the spring of 1969, after many false starts and dead ends, the Edwards team made a slight change in the acidic content of their medium, following the pH of 7.6 which had already proven helpful for fertilizing hamster eggs *in vitro.* "Suddenly," Edwards later recalled, "to our unbounded delight, the sperm started penetrating the eggs."

Of fifty-six human eggs inseminated, thirty-four continued on to maturity. Only seven showed signs of being actually fertilized some thirty-one hours later. And of these five proved abnormal. Two normal fertilizations out of fifty-six! The Edwards team cited the formation of the balloonlike pronuclei and expulsion of the second polar body as their criterion for fertilization. Rather than run the risk of adverse criticism and moral censure as Petrucci had, they decided to terminate all experiments before the zygote divided into two cells, which is for many scientists the decisive mark of fertilization.

The Edwards team offers two theories to explain the high percentage of abnormal developments, five out of seven. First, the eggs they used may have been somewhat degenerated before they were collected from the fallopian tubes or damaged during the collection process. The other alternative is also interesting in its implications. The high percentage of abnormal fertilizations may be perfectly natural and normal. *If so, more than half of all the naturally fertilized eggs abort unnoticed in the first few days after conception.*

This whole process has been pushed one step further with the work of Dr. Cecil Jacobson, assistant professor of obstetrics and gynecology at George Washington University. Dr. Jacobson recently unveiled photographs of sperm penetrating human eggs, the extrusion of the second polar body, formation of the pronuclei and their union. He went another step and examined the chromosomes from the nuclei of cells in the blastocyst, when the egg is in the four-cell stage. But again the experiments were terminated long before full development could be reached.

These experiments with test-tube fertilization of human eggs bring into immediate and sharp focus the basic question confronted by the medical profession and the man and woman in the street when an abortion is suggested. When in the two-hour-long process by which a human sperm adds its hereditary information to that of the egg during fertilization do we *first* encounter a human being endowed with a human life principle or soul and possessing the inalienable right to live? Does this critical "moment" come when the process of fertilization is initiated, or at the end of the union of egg and sperm? Should we draw the crucial line later on, say when the fertilized egg begins or completes its first cleavage division? Or should we place the critical threshold to humanity even later, when implantation occurs? If a human nature is present from the first moment of fertilization, then the experiments of Petrucci, Edwards, Bavister, Steptoe, Jacobson, and many other scientists constitute willful homicide, for in all of these experiments human eggs were supposedly fertilized and then killed. This was precisely the judgment of Monsignor Fausto Vaillanc, the Vatican's press officer when he termed the Edwards experiments "immoral acts and absolutely illicit."

The ethical principles contained in the Hippocratic oath accepted by every medical doctor as a guide for his practice are seriously strained in the questions posed by test-tube fertilization and the possibility of raising a human baby full term in an artificial womb. At present we are faced with the cold reality that none of the human eggs artificially fertilized by scientists outside the womb will develop beyond the earliest stages of fetal life.

In appraising the ethical implications of *in vitro* fertilizations, Dr. Kenneth Greet, a leading Methodist, referred to the criteria proposed by the British Council of Churches six years ago: "Provided the fertilization of the ovum is undertaken in order that it should be transplanted into the lining of the uterus, and provided also no harm is done at that stage which would result in malformation, then I think it is something to be welcomed." But when it came to the possibility of a baby spending a full nine months in an artificial womb, Dr. Greet objected that this would be ethically immoral. In the Roman Catholic tradition, Arch-

bishop George Beck of Liverpool expressed a similar view pointing out that experiments such as those of Edwards are "a travesty
of the natural way in which a child should be born." Many theologians, like Archbishop Beck, see in the Edwards experiment
nothing more than a "totally immoral" extension of artificial insemination. In this context it is interesting to note the parallels
between these ethical judgments and those pronounced a hundred years ago when doctors first proposed the use of anesthetics
in childbirth: to bear a child under anesthesia was a travesty of
nature and a violation of God's punishment in Eden that women
bear their children in pain.

Officials of the Episcopal Church have maintained neutrality
on the morality of artificial-womb babies. Bishop Ronald Williams
of Leicester suggested that the Edwards experiment poses
questions similar to those already raised by artificial insemination. "The church would not want to see any restrictions imposed
upon research into the beginnings of life and it is certain that
the outcome of such experiments will never undermine belief in
God as the creator."

The ethical issues raised by *in vitro* artificial insemination and
the artificial womb are radically the same as those posed by abortion. As such they can be intelligently approached only by the
mutual sharing of views and insights drawn from both the biomedical and philosophical disciplines. The usefulness of such col-
laboration was amply demonstrated in September of 1968, a few
months before announcement of the Edwards experiment, when
Science News carried an informative exchange of views. Dr. Bent
Böving, an eminent embryologist at the Carnegie Institution of
Washington, challenged Dr. Paul Ramsey on the question of
abortion and contraception. Dr. Ramsey, Paine Professor of Religion at Princeton University, had previously urged that we
define contraception "as any pregnancy-preventing measures applied any time before segmentation would occur." His point was
that we should draw the line between morally acceptable contraception and immoral abortion at that point in development when
the organism can no longer divide to form Siamese or normal
identical twins. Dr. Ramsey mistakenly located this turning point
during blastocyst formation, three or four days after fertilization.
In blastocyst formation the solid mass of undifferentiated cells

gradually develops a cavity with a solid knot of cells known as the inner cell mass. This inner cell mass will eventually become the embryo proper while the outer encapsulating layer of cells becomes the nourishing placenta. As this development is occurring, the blastocyst is also preparing to implant itself in the lining of the womb after its trip down the fallopian tube ends. Ramsey's point was that the dividing cells, the conceptus, "cannot be considered to be invested with individuality" while it is still capable of forming identical twins. Hence stopping its development prior to this point is morally acceptable, whereas later interference would not be.

Dr. Böving responded by pointing out that Siamese and identical twins result from a splitting of a single primitive streak, the future spinal cord and central nervous system which begins forming about sixteen days after fertilization. Thus, if we accept Ramsey's argument, it would *not* be an abortion to terminate a pregnancy during the first two weeks of pregnancy. "We have a fellow human being, having claims like our own, *not earlier than* the time when it has been irreversibly settled whether there will be one, or two, or more individuals launched into existence from cleavage of the genotype." Dr. Ramsey is here referring to the old dilemma posed by an acceptance of the infusion of the soul at fertilization, namely, if a fertilized egg divides some days later to form identical twins, are there two souls present at fertilization or does the creator infuse the second soul later on when needed by the second identical twin? Dr. Böving was clearly opposed to solving the question of abortion's morality by extending the meaning of contraception to blastocyst formation. His reason was that while conception usually is considered a synonym of fertilization, the simple union of egg and sperm, this equivalence appeared only in the last two hundred years. Prior to that time, because of little or no information about eggs, sperm, and fertilization, Böving suggests that conception was used in its strict etymological meaning.

If fertilized eggs are experimentally transferred from their mother to a foster mother or culture dish, the genetic mother is no more pregnant than the father. The foster mother becomes pregnant; the culture dish hardly. The deciding factor is the female mammal's conception (L. *con capio:* capture, grasp, take in) of the developing conceptus (the captured object) through the process of implantation.

Implantation begins about a week after fertilization and this, strictly speaking, is the conception of the embryo. In replying to Böving, Ramsey conceded that perhaps two weeks after fertilization, instead of three or four days, should be considered the dividing line between abortion and contraception, but he did not explain how he could still use the term contraception for pregnancy-preventing measures applied between implantation (conception) and the formation of the primitive streak a week or more later.

The complexity of the abortion question today is evident to anyone who does any reading at all. I do not propose to explore that problem in any detail here. However, the basic moral issues raised by abortion closely parallel those posed by *in vitro* fertilization. In simple terms, if a new human being is present from the time of fertilization, the traditional issue of abortion becomes central in dealing with *in vitro* fertilization today. If the embryo becomes truly human sometime after fertilization, then we must agree on when this occurs, at least approximately, so that we can discuss rationally our present halting steps toward an effective artificial womb.

In terms of the preformation and immediate animation theories which have dominated ethical and philosophical thought for the past three hundred years, the answer has been simple, neat, and clear: fertilization is *the* critical threshold. Before fertilization a true human being does not exist; after that event we definitely do have a human being. Contraception may be unacceptable to some as an interference in the reproductive process, but at least it is not abortion.

Where has this dialogue between medical doctors, experimental embryologists, and philosophers brought us in the context of a process view of man? Do we arrive at the same, or a similar ethical appraisal?

Some still hold that fertilization is *the* turning point because once this process is complete the zygote has a complete complement of human chromosomes properly encoded with all the hereditary information required for the development of a new and unique human person.

Others, because of the high percentage of fertilized eggs which abort before implanting in the lining of the womb, place the critical point at the end of the first week of pregnancy.

A third group, because of the question posed by identical twins, argues that a human being is not properly present until about the sixteenth day after conception when the inner cell mass has lost its capacity to split in two to form twins.

These three evaluations, however, do not exhaust the possibilities within a process view of human nature. A fourth alternative has been proposed by Joseph Donceel, a Belgian Jesuit teaching at Fordham University. A philosopher of outstanding reputation, Father Donceel is also the author of a standard college text, *Philosophical Anthropology*. He admits that no one will ever be able to determine for certain the exact moment the human soul is infused or even the approximate time when the embryo becomes human. Nevertheless, he clearly states:

> I feel certain that there is no human soul, hence *no real human being, during the first three months of pregnancy,* and *I consider as safe the norm which applied in medieval Catholic England, where the common law permitted abortion when it was performed before quickening.*

Like fertilization, quickening of the embryo in the womb at the beginning of its fourth month is a somewhat arbitrary norm. It is difficult, if not impossible, for an expectant mother to discern exactly when the embryo in her womb begins to kick and stretch. Yet quickening does mark a definite stage in the development of the human person. "It shows," in Donceel's words,

> that the embryo has definitely *passed from the vegetative to the animal stage,* from the physiological to the psychic level of existence. *The next stage, which is most probably still quite a distance away, is the human stage,* the spiritual level. We are no longer so certain, we may no longer act [to interfere with the process of development].

Donceel maintains that his position is the only one compatible with the centuries-old practice of the Roman Church which after the Council of Vienne in 1312 forbade the faithful by law to baptize any miscarried embryo unless it showed at least some sign or form of being human.

In the fall of 1968 Dr. Donceel presented his views in a formal paper at the opening session of an international conference on abortion sponsored by the Association for the Study of Abortion. At this same session yet another interpretation of human origins was offered by Rabbi Margolies. In even the strictest Orthodox Jewish tradition the fetus is not considered human until it is born. Abortion is permitted because the fetus is regarded as "an integral part of the mother's flesh" and hence under her complete dominion until birth. Jewish practice, the Rabbi recalled, does not permit funeral services for any fetus.

The views of Donceel and Margolies, if accepted, might resolve a sharp dilemma for theologians facing the reality that a good percentage of all the human eggs fertilized abort before going very far in their development. If fertilization marks the beginning of human existence as such, then as Karl Rahner, the respected German theologian, pointed out, countless human beings die long before they ever have a chance to make a moral decision with respect to an afterlife. This would make limbo as populated almost as those other two better known places, heaven and hell. But if the critical threshold is somewhere after the first trimester of pregnancy, then those aborted embryos suffer the same mortal fate as sperm and eggs which never participate in fertilization.

One pertinent question, though, which must be faced is an integration of the views posed by Böving, Ramsey, Donceel, and Margolies with the facts of modern genetics. Is the human genetic code which exists in the fertilized egg not enough in the way of a *preformed* information code to constitute that zygote truly, if not yet fully, human. This approach to a solution may look enticing, but as Dr. Sheldon Segal, a specialist in biomedical studies, noted at the conference, the whole concept of DNA and protein synthesis is based on microbial (bacterial) genetics. Geneticists have not yet shown that this microbial system is *the* system in mammals, let alone in humans, though they commonly function on this assumption. Dr. Segal suggested that basing an argument for theological or moral conclusions about abortion on this flimsy base can be dangerous, blindly walking the plank, so to speak. A variety of human cells have been maintained *in vitro* by tissue culture techniques, but no one would argue that these cells are human in any sense beyond the fact that they each contain a complete complement of DNA and genes for human development.

This appraisal of the origins of the human person alleviates, at least for the next few years, the dilemma faced by scientists who resort to *in vitro* fertilization in their attempts to understand better the complexities of human reproduction and to assist women to have normal pregnancies. In applying his proposal to abortion, Dr. Donceel admits that there is plenty of room for abuse and for selfish (immoral) decisions on the part of the mother. Hence he insists that "although the pre-human embryo (in its first three months) cannot demand from us the *absolute* respect which we owe to a human person, it deserves a very great consideration, because it is a living being, endowed with a human finality, on its way to hominization," to becoming fully human. Therefore, he argues, abortion in the first three months of pregnancy should not be permitted unless "very serious reasons exist." We can extend this philosophical judgment into the area of artificial wombs and *in vitro* fertilizations with very minor adjustments. In the next few years we are not likely to face the question of risk in terminating the life of an unborn human fetus beyond the three-month period before quickening. Hence our ethical question is whether the experimental embryologists and physicians have sufficiently grave reasons for triggering the test-tube fertilization of a human egg and allowing it to begin development in an artificial womb when they know they cannot carry it to full-term development.

Why are the scientists pursuing this work with test-tube fertilization and artificial wombs? Are they not content with nature's tried and true method of reproducing the human species? Each scientist obviously has his own reasons.

A French biologist and winner of the Nobel Prize for his work with Jacob and Monad on the genetic switch mechanisms which underlie development (the Operon theory), Etienne Wolff, is very much concerned with research in test-tube fertilizations. As director of the Laboratory of Experimental Embryology associated with the Collège de France, Wolff has highlighted some advantages we can expect as this technology progresses. It may make it possible for doctors to save fetuses that would otherwise die, to stimulate the intellectual and physical qualities of the child as it nestles in its glass womb, to repair congenital malformations, and to immunize infants against infection by prenatal vaccination. Work is already progressing on developing tolerance to foreign tissues and blood by prenatal immunization of the fetus so that it will be able to accept transplant grafts and transfusions later in life without danger of rejection.

Ectogenesis, artificial fertilization followed by nine months in an artificial womb, would allow a woman prone to miscarriage to have a child from her own egg and her husband's seed. If fertilization were accomplished through normal intercourse, could anyone object on moral grounds if the doctor collected the embryo before implantation and transferred it to a glass womb, bypassing the dangers of miscarriage? Some might consider this just a very early Caesarean section. If a woman's reproductive tract is so hostile to her husband's sperm that she cannot conceive but can carry a child, should the doctor try *in vitro* fertilization with her egg and her husband's sperm and then implant the embryo in her womb?

Nino Lo Bello has quoted Dr. Petrucci on the rationale behind his experiments.

One of my aims is to help women have babies, for I have been upset by the large number of women giving birth to still born children, especially at their first pregnancy. Thus my research was directed along humanitarian lines, guided by the Christian principles I have practiced since childhood. I love mankind. If a wife should lose a baby on which the hopes of herself and her husband have been centering, this is a human tragedy. That I should be denounced for my experiment is a great personal blow, for I am a scientist dedicated to uncovering those mysteries of nature that God is prepared to reveal to us.

Once doctors can routinely fertilize a human egg *in vitro* and observe its development in detail, a whole new field of medical research opens up. Fetology is already an important field of medical research and it is likely to become one of our prime concerns tomorrow. Obviously, as man gains more and more control over his reproduction and with the population explosion reducing the number of children per family, it will become vitally important to parents and physician alike that each child that is conceived receive the very best possible treatment and care during its embryonic life. An effective artificial placenta and womb would make it possible for scientists to determine when different enzymes become functional, what these enzymes are, and what specific developments they trigger or their absence prevents. The nature and origins of various biochemical abnormalities could be ascertained and perhaps prevented. Major advances in effective and safe contraceptives would also be likely.

Test-tube fertilizations offer another avenue to the selection of sex which we spoke of in our first chapter. At Cambridge University, Dr. Edwards has worked with Dr. Richard Gardner on this from a postfertilization vantage point. After gathering embryos from the fallopian tube of the rabbit before implantation, they found they could cut into the blastocyst cavity without damaging the inner cell mass which is the embryo proper. By withdrawing a few cells floating in the fluid of the cavity and examining these in a chromosome smear or karyotyping, they can learn the genetic sex of that embryo. Besides the XX and XY chromosome differences, body cells of the female contain a dark-staining chromatin mass in the nucleus known as the Barr-body. This mass, scientists believe, may be a shriveled, inactive X chromosome, since only one X chromosome appears to function in each cell. Males seldom show this Barr-body in their cells. After the embryos have been sexed and the tiny wounds have healed, those of the desired sex can be transplanted back into the rabbit. In all cases save one normal fetuses developed. The one headless fetus might be blamed on faulty technique with some of the inner cell mass escaping through the slit. In every case the embryo was of the predicted sex. While almost infallible this technique does involve aborting those embryos not of the desired sex, and many undoubtedly would find this objectionable. Even Dr. Edwards, speaking at an international symposium in Berlin, admitted this. "Naturally, we would not use this procedure on human embryos. You can be sure that we have other ideas, but their application is somewhere in the future."

British scientists responding to queries from reporters at the time of Edwards' experiment were quite vehement in denouncing the phantom specter of test-tube babies as "pure science fiction." This seems a bit pessimistic and negative in terms of the present state of technology. While it is undoubtedly true that the feasibility of raising a child from fertilization to decantation is not likely for a decade or two, it would be foolhardy to ignore the possibility just because it upsets us emotionally. One of the biggest problems with theologians and moralists, as well as the man in the street, is that they almost always wait till something becomes a reality before *reacting*. I use reacting here instead of responding, because it is a much more appropriate description. There is no dialogue in these situations, for instance, Monsignor Vallainc's dogmatic and emotional reaction to Edwards' experiment.

One is tempted to wonder, though, whether an intelligent dialogue is yet possible between the experimental scientists and the official spokesmen for Christian morality. Dr. John Marshall, a British physician and member of the papal advisory commission on birth control, has conceded rather candidly that "the church's theology is not sufficiently developed to cope with" these new developments. Almost without exception, the various Christian denominations have retained the classic views of human sexuality, woman and her place in society, and the nature and function of sexual intercourse, so that they now find it very difficult, if not impossible, to understand or appreciate either the new technologies of human reproduction or the new cultural perspectives these are molding in the minds and hearts of young men and women today. Before we can have in-depth dialogue on the issues of new modes of human reproduction theologians, philosophers, and students of ethics will have to drop the outmoded and antiquated concepts of Plato, Aristotle, Augustine, and others whose views of sexuality and love are permeated, in varied but deep ways, with a fixed world image in which human sexuality is accidental to man's nature and unchanging in its reality. The ubiquitous subterranean puritan-gnostic dualism which relegates human sexuality to the dirty and shameful must also be

scrapped once and for all. To achieve this new perspective philosophers and theologians must accept the process view of man
with all its implications, viewing man as a constantly evolving
sexual person and human society as a community of sexual persons who express their sexual personalities in a pluralism of
acceptable forms. Without this radical shift in perspective, I am
certain no fruitful dialogue will ever develop between the specialists who must in some way reflect and respond to the masses
of mankind.

In discussing the ethics of *Man's Intervention in Nature*, Owen
Garrigan has pointed out that the decisions and judgments of
medical ethics are molded by and reflect the ethics of the man in
the street. In this context one is tempted to wonder about the
reaction of any person faced with the decision to abort a two-
month-old embryo after he has studied Lennart Nilsson's remarkable color photographs of the embryo in the uterus. If you
had watched an unformed mass of cells slowly take on all the
features of a perfectly formed human in miniature, hardly an
inch long from crown to rump yet complete even down to tiny
fingernails and embryonic teeth buds, could you make the decision to pull the plug on the artificial womb? As time passes, it
becomes more apparent that fewer and fewer young will be
conceived, but their conception will be more and more a deliberate and loving decision rather than the chance result of
emotion.

The fact that man is the only mammal to walk erect on two
limbs has raised havoc with his mode of reproduction. With the
other mammals, the process of giving birth is simple and natural
and safe both for mother and offspring. The other mammals do
not run the risk of toxemia or hemorrhaging to death as a result
of birth, nor do the mothers bear their offspring in pain. Man,
however, has chosen to walk erect on the face of the earth. To
adapt his skeleton for this new and unique posture, thousands
of years of natural selection worked to modify an arched spinal
cord into an S-shaped support. His pelvic girdle was strengthened to bear the weight of all his body organs. And in the process
his internal organs were shifted around to new positions. At the
same time his ancestors slowly gave birth at earlier and earlier
stages of development. The brain was becoming crucial in man's

emergence, and birth must occur before the fetus's head becomes too large to pass through the pelvic girdle. All these changes, and others, may well account for the fact that the human female is the only mammal to bear her offspring in intense pain and suffering, running the risk even of death. In the seventeenth, eighteenth, and early nineteenth centuries upward of 20 percent of mothers delivering in some hospitals were dying of puerperal infections! If man's upright position has created such risk in human childbirth, might it not be reasonable to see in the development of an artificial womb the possibility of rectifying a dangerous side effect nature has not been able to take care of? Birth is a vital biological process, for both mother and child. For life to continue it must be as safe as possible and as functional as possible. If man cannot have a normal painless and riskless birth because he continues to walk erect, then perhaps the solution is the artificial womb.

Two questions must be answered though before we make that decision. Can we produce a perfect imitation of the human womb without in the process testing its preliminary stages on guinea-pig human fetuses beyond the three-month age? These fetuses would run the high risk of ending up as monstrosities that would be aborted. We have already seen the complications and obstacles scientists faced, and overcame, in simulating artificially the rather simple environment required for fertilization of a human egg *in vitro*. The nine-month environment of the human womb is far more complex and intricately balanced. Trace elements and other components of that environment may easily be overlooked in developing an artificial womb; their absence may only be noticed, disastrously and pathetically, sometime after birth. We are venturing into a jungle and maze of hormones, enzymes, trace elements, and the like such as has not faced men of medicine perhaps in their entire experience. Each trace element, each enzyme, each hormone must appear at just the precise moment in the nine-month schedule; each must be in its specific concentration; and each must then taper off or disappear at just the right time to trigger normal development. Can medical science hope to turn the trick? Many medical men and scientists believe we can. Others do not. Do we risk the venture?

The second question is perhaps just as complicated and unanswerable—at least today. What might be the psychological effects of raising a human fetus for nine months in a womb of glass and steel? We know of intrauterine conditioning and learning, elementary as this may be. If the nervous tensions of a high-strung mother can influence a child in the same direction, at least for some time after its birth, surely the warm exchange of mother and child is part of its normal development. The natural womb takes the child through cycles of waking, bouncing around on the way to the supermarket, sleeping, rest, and all the turmoil of a day in the womb around the house. After quickening there is the continual testing of the womb, the fetus kicking against the mother's resilient uterus, and similar exchanges and dialogues. Can we imitate these? Perhaps they are not essential to a normal development. But perhaps, too, they are. Nurseries and orphanages have learned of the newborn's need for cuddling and for the reassuring beat of the mother's heart—even if it be the substitute beat of a metronone! Do we program such stimuli into our artificial womb also?

Psychologist Harry Harlow has carried out some instructive experiments with imprinting infant rhesus monkeys. Isolated immediately after birth, he conditioned some to regard as their "mothers" contraptions made of wire and foam rubber covered with terry cloth and topped with wooden heads. A feeding bottle with nipple completed each substitute mother. This artificial conditioning had an interesting side effect. In later life monkeys raised with dummy mothers failed to develop normal patterns of sexual behavior. Females raised this way would not mate unless they were with a very experienced male. And when they gave birth, they evinced no maternal behavior at all. The effects of the dummy mother can be reduced or overcome if the infants are given an opportunity for sex play with other little monkeys. This experience with the postnatal effects of an artificial mother should raise some serious questions about the psychological side effects of any artificial womb we create.

In Aldous Huxley's *Brave New World* the young students following the director of the Central London Hatchery and Conditioning Center were shocked when Mustapha Mond, one of the World Controllers, lectured them on the glories of their utopian world. They had never heard such language before, and fascinating as Mond's lecture was, it embarrassed the young students no end. His description of the horrors and disgusting features of old-fashioned family life was not that bad, but the words he used, those six-letter words like *mother*, *father*, and *infant*, they never expected to hear from a man as eminent as Mond. These words were to say the least impolite in the presence of young impressionable students like Lenina Crowne, one of the Alpha workers at the Hatchery. They were grotesque as they recalled that barbarous period in man's ancient history when young were conceived blindly and born in pain and blood, after animalistically weighing down a woman's belly for nine months.

The tour included the Organ Room where selected isolated ovaries and sperm produced germ cells in their culture bottles, the Fertilization Room where eggs and sperm could be matched at will to produce whatever social type was currently needed, the Bottling Room and the Embryo Store where the bottled embryos spend 267 days before moving on to the Decantation Room. In the Embryo Store, Alpha workers like Lenina maintained a rigid control of the environments in the artificial wombs, applying the proper hormones and drugs, perhaps even some minor surgery to tailor each product to specification. The social predestinators from the College of Emotional Engineering saturated the psychic experience of each embryo to provide just the right number of gamma-minus machine minders who abhorred reading and thinking, Alpha-plus psychologists to brainwash the populace with 6,400 repetitions that made a truth, and other useful forms of men.

Inconceivable in 1932? Yes. Inconceivable in 1830? Certainly! But inconceivable tomorrow or by the year 2000, only thirty years hence? Not so improbable, perhaps. Certainly not impossible.

In pioneering the question *Can Man Be Modified?*, Jean Rostand asked in 1959 whether future generations of men may not view decantation for an artificial womb as natural as we now view childbirth under anesthesia and a caudal injection? Just as some women today prefer natural childbirth while others will have nothing but delivery with a caudal anesthetic, I suspect that mothers in our next generation may show an equal divergence of reactions should an artificial womb become available.

Some, undoubtedly, would consider nothing except the old-fashioned way their mothers did it. But others may agree with Rostand and decide to bypass "the animalistic burden of pregnancy" to decant their baby at the local incubator.

THREE

EGGS
BY THE GROSS
Want to Be a Mercenary Mother?

Have you ever heard of a "supermother"? Or of a "superpregnancy"? I realize most people react very skeptically when they hear the word "super" used today, but in this case the term is very accurate and quite appropriate in describing some actual accomplishments of experimental embryologists.

One of these scientists, a young research technician at one of the many pharmaceutical houses which surround the Madison campus of Fairleigh Dickinson University, visited me just about the time I was halfway through the rough draft of chapter two. He was interested in pursuing a graduate program in biology and perhaps doing his thesis research under my direction. In the course of the conversation I learned that he had a special interest in experimental embryology, specifically in the artificial induction of ovulation by hormone injections. As we talked, he casually mentioned that he likely held the world's record for superovulation of the rabbit. In the course of one of his experiments at work he had injected a doe rabbit with a combination of female hormones normally associated with ovulation. The shocked rabbit responded to this physiological overdose by releasing not the usual eight or ten but well over a hundred eggs. The doe was then artificially inseminated. When she was definitely pregnant, the technician sacrificed her for observation. With 135 tiny embryos developing in her womb, this doe rabbit certainly would qualify as a "*super*mother," even though she could not possibly have carried the 135 fetuses full term and delivered them normally.

As a research technique superovulation has been around for thirty years. The first reported supermother was a rabbit, injected with hormones and artificially inseminated by Dr. Gregory Pincus in 1940. Two years later Dr. A. S. Parkes and his co-workers at London's National Institute for Medical Research repeated the success of Pincus with rabbits, and added sheep to begin a list of prospective supermothers the next year. In 1944 Dr. John Hammond superovulated some cows. In 1949 goats joined the growing list after an experiment by Dr. A. J. Folley and fellow researchers. Rats were added in 1950 by Dr. C. R. Austin at Tulane University. Mice came along as supermothers in 1953, monkeys in 1957, and hamsters in 1962.

At first this work seemed to be what scientists like to call "basic research," that is, research that is undertaken simply because a question arises about the nature or effect of some technique which at the time appears to have absolutely no practical application outside the plain laboratory experiment and the acquisition of another bit of information. But, as frequently happens with basic research, several applications soon emerged.

Dairymen and sheepherders quickly became interested in superovulation as a possible way to increase twin offspring in both cattle and sheep. In the course of nature, cows and ewes occasionally give birth to twins, but if the farmer can induce his animals to have twins every time, the production of veal and lamb could be greatly increased and its price reduced, much to the delight of gourmet cooks and housewives.

Butchers sometimes have a partiality for young hogs as a prime source for choice cuts of pork. This can create a dilemma for the hog raiser. If he sends all his sows to the slaughterhouse before they mature sexually and produce at least one litter, he has to replenish his stock by purchasing very young porkers. If he waits till after his sows mature and have a litter or two, their price drops at the slaughterhouse. Experimental embryologists got the hog raiser off both horns of his dilemma in the 1950's. Dr. William Marden tried to superovulate a sexually immature calf, and surprisingly he was successful. In 1959 several researchers tried to superovulate immature rats: Gregory Pincus, E. S. Hafez, R. G. Edwards, and others had success. The next step was to try it on pigs to see if, with hormone injections and artificial insemination, a young sow could be induced to produce a healthy litter long before reaching natural sexual maturity and just before being hustled off to the slaughterhouse to be made into tender pork chops. Success! The butchers were happy, the hog raisers were happy, and the housewives were happy.

Then a very practical question came up: if veterinarians can induce a six-week-old calf or a very old cow to ovulate or superovulate with hormone injections, can we solve certain cases of human infertility the same way?

In the mid-1960's research moved ahead very rapidly on this project. Rebound effects were reported when women went off the contraceptive pill. With the simulated pregnancy induced by the pill suddenly halted, some women responded with an almost immediate ovulation, often not just of one egg, but of several. In one case a woman had stillborn octuplets after using Perganol, one of the "pills." Various combinations and dosages of hormones were tried, with a variety of results. Eventually, the rebound effect was controlled and in many cases doctors were able to induce single ovulations in previously childless women. A hormone-combination pill, such as Clomid, taken for three or four days, at any time in the monthly cycle, often triggers ovulation of a single egg within a week. Thus for some women the burden of a childless marriage has been resolved. But the results are not always predictable, as this technique is still quite experimental. In October of 1968 Mrs. Sheila Ann Thorns, of Birmingham, England, upset some statistical odds in her response to a gonadotropic hormone. Childless at the age of thirty, she gave birth by Caeserean section and two months prematurely to sextuplets, four girls and two boys. Statistical odds for this phenomenon are normally about one in every 3.2 million births.

Hormone-induced quintuplets were born about the same time in Sweden, New Zealand, and New York City.

This is all very fine but who really wants an instant family via sextuplets? And what can you do with a superpregnant rabbit, expecting centuplets, even when this is in the laboratory and not in your son's hutch in the backyard?

One practical outcome of superovulation and superfecundation has emerged from the work of Dr. E. S. E. Hafez at Washington State University. In 1964 Dr. Hafez reported his first experiments on the effects of overcrowding in the womb of a superpregnant rabbit. The rabbit's womb is designed by nature to carry between six and ten fetuses comfortably at a time. Thousands of years of evolution have made it extremely effective in handling this normal physiological load. But what happens when you overload the system? Does it break down completely when overloaded with a hundred or more fetuses at one time? Or does the system compensate by somehow reducing the load back to normal? This question is closely linked with another: why do some women conceive what appears to be a normal fetus and then abort it spontaneously? Why do so many embryos die soon after conception?

Fetal death is a no-man's land, filled with countless unanswered questions, but Dr. Hafez is convinced, as are many other scientists, that the answer probably lies in the placenta. To test this theory he superovulates cows, sheep, and rabbits. Following the artificial insemination, multiple pregnancies place the placenta under terrific strain, pushing it often to the limit where compensation or breakdown occurs. By studying the placenta in this stress situation, Dr. Hafez hopes to learn what process goes on in a cow's uterus which reduces an overload of eight or ten fetuses to a more manageable one or two. Does the overloaded placenta simply select at random one or two embryos and eliminate the others by shutting down their nourishment? Scientists such as Hafez and Krantz, whose placenta simulator we mentioned above, believe that we may soon gain control over placenta malfunctions so that many women who conceive normal fetuses will be able to carry these full-term through placental medication. We may even be able to improve on the functioning of a normal embryo.

The pioneering experiments of Drs. Walter Heape and F. H. A. Marshall at the turn of the century did not lead to any practical application for superovulation until about fifty years later when the Russians became very active in this research field. Early indications of possible applications came from closely related research in 1949 and 1951 when Dr. Raymond Umbaugh reported his attempts with ovotransplants, the transferring of a fertilized egg from one cow to another. In 1953 Dr. E. L. Willett and his associates at the University of Wisconsin and the American Foundation for the Study of Genetics reported their successful efforts in transplanting individual embryos from the natural mother to a substitute or foster mother.

Around the same time Dr. L. E. Rowson began gathering a team of specialists, biologists, and veterinarians at the Agricultural Research Council's estate near Cambridge, England. Before long he was joined by Drs. C. E. Adams, G. L. Hunter, and R. L. W. Averill. In 1957 Rowson and Adams reported success in transplanting fertilized eggs of sheep from one ewe to another. Nishikawa, Horie, Surgie, Onuma, and Niwa persuaded the goat to follow suit and permit her embryos to be transplanted in 1963. Before long ova transfers, or inovulation as this technique is now known, were being tried by Hungarian, Polish, Italian, Spanish, Russian, Japanese, British, French, and American scientists, hard at work improving the techniques.

In short order scientists were crossing the species and breed lines to test the limits of inovulation. Under Rowson's direction a black Friesian cow served happily as a substitute mother for a Hereford calf with its white face and reddish-colored body. Likewise an albino rabbit served as substitute incubator for a pure black offspring.

In 1962 the ubiquitous and versatile rabbit once again came to the aid of the experimental embryologist. The plan devised by Rowson and his co-workers called for two Dorper ewes in South Africa to give birth to purebred lambs of the Border Leicester strain from England. This could be accomplished by fertilizing, in South Africa, Border Leicester eggs with sperm from the same breed, both of which had been shipped frozen from England and then implanting the product in the wombs of Dorper ewes. In theory this approach was excellent, but in 1962 the technique of freezing eggs was still not perfected. The solution proposed by Rowson involved fertilizing Leicester eggs in England and then shipping them to South Africa for transfer to the substitute mothers. The *Journal of Reproduction and Fertility* carried the full story of what has become the basis for much speculation and countless thought-provoking questions: the possibility of using an animal of another species as temporary or full-term incubator for a human embryo. "The Successful Long-distance Aerial Transportation of Fertilized Sheep Ova" was the title of Rowson's paper in that journal. It detailed the technique by which the eggs were fertilized in England and placed in the fallopian tubes of a rabbit. Nestled snugly in their cozy incubator, the fertilized eggs continued to grow and differentiate while their substitute mother was carefully crated and loaded aboard a jet liner for South Africa. On arrival at their destination, the tiny embryos were surgically transferred to two Dorper ewes which had been tricked physiologically by hormone injections into thinking that they were about to become pregnant. Very quickly the expectation became a reality. This science-"fiction" story came to a successful climax when the ewes delivered two perfectly normal lambs.

Surrogate mothers, substitute mothers, mercenary mothers—however you choose to describe them—were no longer mere grist for speculations and fantasies of the science-fiction writers. They had now become part and parcel of man's techology of reproduction, one of the several methods available to us in our exploding ability to manipulate and control our own reproductive behavior.

Scientific experiments do not suddenly sprout in full bloom out of the air. They undergo a long gestation period, sometimes centuries in length, before they come to full birth. Thus by the time an experiment is successful scientists often have worked out practical applications. In some cases, as with Rowson's experiment, both theory and application are conceived in detail long before the technique itself is successfully worked out.

The theory behind the expected feasibility of embryo transplantation rests on the lack of the rejection response in the womb. Knowing what we do about the body's immediate defense reaction to any foreign tissue, one might expect the alien tissue of an embryo to trigger an immune rejection by the mother's body. However, if this happened, no woman could ever have a normal pregnancy. Fortunately, the processes of natural selection and evolution have adapted the womb in some way so that it is not subject to the triggering effect of alien tissue. Interestingly enough, a piece of fetal skin is often attacked and destroyed when transplanted to the mother's arm while the fetus remains safe and unharmed in the same mother's womb. Apparently the tissues of the womb and particularly the placenta act as a protective barrier, preventing the mother from developing an immune rejection. Occasionally, this barrier is not completely effective, as for instance when fetal Rh positive blood leaks through the placenta and triggers an immune response in the Rh negative mother. Ordinarily the mother and fetus can coexist without any difficulty despite the fact that they have different blood types. This background led scientists to suspect that if any transplant would succeed it would be the transplant of an embryo from one womb to another.

Even at that time, the possible applications for inovulation were fantastic. In animal husbandry alone—the human implications we will deal with shortly—some very important advances could be made both in terms of cost reduction and in genetic improvement of herds. Dairy- and cattlemen have for years been disturbed by the inefficiency and limitations imposed by nature on their hopes for improving their herds. A prize cow can only mother ten to twelve calves in her lifetime. Champion cows, like champion bulls, are not that common. But at least in the case of bulls, frozen sperm used in artificial insemination programs permits half of the rare genetic combination which makes for a champion to be shared with perhaps fifty thousand offspring a year. The problem is to find fifty thousand champion cows for this prize bull to inseminate! The ratio of fifty thousand to one is hardly appealing to cattlemen. Even superovulation is useless in this case without the ability to transplant some of the champion offspring to substitute mothers. If it were possible to transplant embryos, then a champion cow could be superovulated, inseminated with frozen semen, and the tiny embryos collected before they implanted in her womb. A combination injection of PMS (pregnant mare serum) and serum extract from a pregnant woman will cause a cow to ovulate upward of forty eggs at one time. This could place thirty or forty prize embryos in the veterinarian's syringe for transfer to the wombs of thirty or forty healthy but genetically inferior cows. There are always plenty of second-rate cows around a herd, so that with superovulation and inovulation with a single prize cow and frozen semen shipped in from elsewhere a dairyman can end up with three dozen prize calves instead of one or two at the end of a single breeding season. From then on the quality of his herd would be far better than he could ever hope for relying only on limited finances and Mother Nature.

Naturally inovulation did not create a second Eden for the cattlemen. There were some drawbacks and problems involved in the new technique. With cows, for instance, it is sometimes difficult to get the follicles containing the extra eggs to rupture and release their germ cells. Nature has conditioned the ovary to mature and release one or two eggs in each estrous cycle. When man intervenes to trigger the maturation of a couple of dozen eggs at a time, the ovary tries to continue its old pattern of ovulating only one egg even though many more are mature and ready to be released. The solution to this problem may be in the relative proportions of the two basic hormones used. Also there is a certain limit to how many eggs you can get from a cow. If too many eggs are involved in one superovulation, their capacity to be fertilized declines.

In the few short years since Rowson's success with sheep embryo transplants the techniques have greatly improved. Originally the recovery of the fertilized egg and its implantation required surgery, an operation quite similar to a Caesarean section. Today, as a result of research in 1964 at the Livestock Industry Experimental Station operated by the Japanese Ministry of Agriculture, the surgery has been eliminated. Fertilized eggs can now be washed from the fallopian tubes of the donor cow and injected into the host womb with a double-barreled syringe. One barrel distends the womb like a balloon with carbon dioxide while the other barrel inserts the fertilized egg at just the right moment.

Undoubtedly the leading proponent of superovulation, inovulation, and embryo transplantation is a former associate of

Parkes and the Cambridge team, Dr. E. S. E. Hafez, chairman of the Department of Animal Sciences at Washington State University. An Egyptian by birth, Dr. Hafez has plunged into inovulation with all the gusto of a man convinced to the core that his project holds tremendous promise for the future good of mankind.

Albert Rosenfeld, former science editor for *Life* magazine, tells of ten symbolic vials supported in a sponge-rubber base and petri dish which Hafez often points out to visitors to his laboratory in Washington. Each rubber-capped vial bears a neatly typed label: swine, sheep, cattle, rabbit, man. Inside each is some colored liquid. For Hafez this simple arrangement of glass tubes symbolizes animal reproduction in the future, completely separated from sexual intercourse, bypassing or seriously modifying the traditional monogamous couple marriage of humans, removed from its familial and perhaps even from its maternal setting, and occurring across the far reaches of both space and time. A dozen or two vials containing frozen or fresh zygotes could supply the basic mammalian, including human, population for a new world. Shipped to another planet, they would require only insertion into the appropriate artificial wombs to be born. With a scientist or two along to supervise and care for them during their infancy, the problem of dead weight in space vehicles could be completely bypassed. Hafez has asked, if we miniaturize the electronic components of our spacecrafts, why not also miniaturize their biological components? Dr. Hafez' work has been heavily subsidized by the National Institute of Health, and he sees no reason, with the space program now moving into plans for a landing on Mars, why basic research with zygote storage, inovulation, and the artificial womb should not be a part, and an essential one, of that program.

The balding, slightly stooped professor has made a speciality of inovulation over long distances. After inducing a cow to multiple ovulation he uses artificial insemination, either with frozen or fresh semen, and gathers the eggs after fertilization occurs. The tiny embryos are then implanted into the womb of a pseudopregnant doe rabbit where they can survive happily for a fortnight while winging their way across oceans to some developing nation. Rabbits only ovulate during coitus, so in this case the doe is mated with a sterile buck and injected with hormones so that she thinks she is pregnant. When properly timed, this pseudopregnancy brings her womb to just the right point of the cycle for successful implantation. Once the substitute mother has arrived at her destination with her bovine passengers, the veterinarians sacrifice her for the cause and transfer each small embryo to a waiting bovine foster mother. The substitute mothers then switch from a hormone-induced pseudopregnancy to an actual pregnancy; the 280-day gestation goes ahead without hitch. A hundred prize calves, offspring of pedigree champion parents somewhere on the other side of the earth, grow to maturity in the wombs of their sturdy but genetically second-rate foster mothers. Rather than ship a hundred pedigree calves or lambs by air freight, we can now ship one mother rabbit by air parcel post at a hundredth the expense and inconvenience.

Could this technique of embryo transfer be applied to man? Neither Dr. Hafez nor the Cambridge team of Edwards, Steptoe, and Bavister see any reason why it could not be used in human reproduction. As a step in this direction Drs. Steptoe and Clyman have already tried transferring human eggs and semen from donors to the fallopian tubes of women scheduled for hysterectomies and similar operations in the hope that the eggs might be fertilized and start developing before the operations. The results were negative. As of early 1969 Steptoe indicated that the Cambridge group had not tried to transfer already fertilized human eggs into a host patient. However, he did predict that this team will be ready for such a venture within a year or so. A survey article in *Medical World News* quoted Dr. Steptoe's reasons for proposing such a venture: "If you could see, as I do every day, the anguish of couples who want babies, you'd want us to do everything possible to help them."

What about the possibility of a subhuman surrogate mother for a child? If a rabbit can serve as a temporary incubator for a hundred calves, might she also serve as incubator for a human embryo during the very early stages of its development? Dr. Edwards is convinced that there is no medical or biological reason why such a transplant would not work. The transplanted fetus is indeed an allograph, a foreign tissue, but as noted above, the uterus for some unknown reason is not triggered to an immune response by the embryo. In fact, Edwards' earlier experiments with implanting unfertilized human eggs into a rabbit foster mother makes him very optimistic about this animal providing an excellent nursery for a very young human being. Before the Cambridge group attempts a zygote transplant between two women, they first plan to put human embryos into such furry four-legged incubators. On February 26, 1970, Dr. Jack Cohen, senior lecturer in embryology at the University of Birmingham, England, suggested that "host mothers" may soon be paid to give birth to babies conceived in test tubes from the eggs of women unable to have children. His comments were part of the public controversy over Mrs. Kenneth Allen's attempt to have a child by embryo transplant. Mrs. Allen, thirty-four, has been childless for seven years of marriage because of blocked oviducts. Her husband had fertilized her eggs in a laboratory experiment in Edinburgh and the doctors planned to return the fertilized eggs to her womb within a few weeks, thus bypassing the blocked oviducts.

Many of these pioneering experiments will undoubtedly fail before the scientists solve all the technical problems, a point which raises again the specter of abortion. The solution proposed by Dr. A. S. Parkes and by theologian Paul Ramsey that we should consider a true human present only after implantation occurs or after the embryo can no longer divide to form identical twins is of little help here, for implantation would occur in the rabbit's womb and termination of the experiment would come about the time the primitive streak or primordial neural tube is formed. It might be interesting to qualify Parkes's suggestion and limit conception to actual implantation in a *human* womb. But what does this imply for implants raised in a subhuman or artificial womb? Donceel's contention that we cannot consider a true human to be present until some undetermined time after quickening occurs would not rule out these experiments with subhuman incubators on the grounds of abortion, though one might not judge the reasons proposed by Edwards and Steptoe serious enough to justify the venture. Women with blocked oviducts are the guinea pigs for many of these experiments. Often these women are so desperate to bear a child of their own in their own womb that they are more than willing to cooperate in the pioneering phases of inovulation. With all the orphans around this earth one wonders about the female (and male) psychology of this reaction. How much of this seemingly natural desperation is culturally conditioned and how much of it is truly a part of woman's innate instinct and psychology? Is it a natural, normal or balanced emotion that prompts some women to go to any limit to bear her own child? Can anyone answer such a question today?

Even on the purely theoretical level the question of human motivation is difficult, if not impossible to decipher. Equally impossible is an answer to that basic question of when an embryo becomes a human being. Obviously, in both situations we can no longer hope for a simple cut-and-dried, categorical answer which will apply to all cases, an answer which everyone will accept. Among the vertebrates and especially among the mammals and primates biological evolution has geared thousands and millions of years of effort into the evolution of a reproductive organism. This has involved much more than a mere anatomical system. It encompasses the complexities of hormone-conditioned psyches, behavioral patterns and instincts, and countless other factors. Generally speaking we can say that the female mammal is geared for reproduction. The question facing man and woman today is to what extent can we transcend this biological-human function, and what effect is this likely to have on human relations? On the question of human origins it is likewise obvious that we can no longer turn to the simple answer of ensoulment at fertilization. Among the solutions surveyed here my own impression is that civilized Western man will gradually opt for some answer along the lines suggested by Joseph Donceel. To my mind his suggestion contains the best philosophical basis and agreement with our biomedical knowledge of prenatal development.

In the first chapter we spoke of AI, artificial insemination, and AID, artificial insemination with a donor's semen. We can now complicate our acronym dictionary by adding artificial inovulation to the possible meanings of AI and artificial inovulation with a donor's germ cell to those of AID. In the latter case, however, we should specify whether only the egg, only the sperm, or both germ cells were donated.

When a childless couple adopts a child, they do it out of love, and the bond between the infant and his new "social" parents is a strong one, even though it is not genetically or biologically grounded. What happens to this familial bond in the case of inovulation which in essence is a prenatal form of adoption? How important are the differences, social and psychological, between postnatal and prenatal adoption? Is it worth the effort and risk? Will it make for a stronger mother-child relationship than simple postnatal adoption? Will it make a significant difference whether a barren woman prenatally adopts and carries a child conceived from the union of a donor's egg and her husband's sperm, or a woman with a blocked oviduct conceives in a test tube with inovulation following, or a woman prone to miscarriage but fertile has her ovum fertilized *in vivo* or *in vitro* with her husband's semen and the resulting embryo transferred to the womb of a substitute mother? We are aware today of some of the psychological and biochemical conditioning that goes on between the biological mother and her fetus during the nine months of prenatal life. To what extent might this create problems for inovulation pregnancies? On another plane, might inovulation resolve or at least alleviate a sterile woman's psychological or marital unsureness? If, on the other hand, a woman decides to have an abortion, should the doctor recommend she donate the fetus to those laboring to develop an effective artificial womb? Or might he recommend an embryo transplant to a woman who really wants to bear a child of her "own"?

Some of the questions we raised earlier in conjunction with artificial insemination arise here in a slightly different context and to my mind with more serious implications. For instance, should the substitute mother remain anonymous to the adopting couple, or would it be better if the prospective parents knew and chose her personally? Should she be a relative? If you were the husband, would you object if your wife suggested her sister carry your child? If you were the wife and unable to carry a child, would you object if your husband's sister offered to carry your baby? Decisions on questions like these would likely depend very much on individuals and their own unique situations. In any event we should look carefully at the relationship a mercenary mother is likely to develop with the prospective parents, if she is known to them. We should seriously consider what effect known mercenary mothers might have on the traditional monogamous family structure. Certainly, the use of mercenary mothers and their possible prevalence in the coming generation raises serious questions about the exclusivity of the couple marriage which heretofore has been based very much on sexual (genital) fidelity. With the appearance of surrogate mothers on the scene are we perhaps evolving toward a broader conception of the family circle?

Once again the classic meaning of terms like mother, father, parent, and our whole conception of motherhood takes on new meaning. Should we now begin, both in ordinary conversation and in legal parlance, to speak of genetic mother—the donor of the egg; natal mother—the female who carries the child; and social mother—the female who raises the child after birth? Similar distinctions might be applied to the male side of human reproduction. But then who is the natal mother when a subhuman foster mother carries the child to birth? A cow? How would you explain that your grandchild has a simian natal mother?

We have had enough experience with unwed mothers and the adoption of their children to realize that many problems we encounter there will be carried over and intensified in the case of surrogate human mothers. What if a married couple arranges for a substitute mother, supplies the egg and sperm and nine months later finds out the foster mother is unwilling to give up the newborn child? Does the married couple have a real claim to that child, their genetic but not natal offspring? How binding legally would a certified contract or agreement be in this situation? What happens if the substitute mother changes her mind about carrying the baby and aborts deliberately after the embryo is transplanted to her womb rather than give up the infant? Can the genetic parents sue her for damages? What might the emotional and legal results be if the surrogate mother takes some drug, prescription or otherwise, during the pregnancy and *inadvertently* causes a malformed baby to be born? How would you feel if your sister or some anonymous foster mother were carrying your baby and the doctor came into the waiting room to inform you that she had just delivered a thalidomide infant. Some cases might not be nearly that clear cut. Researchers at Columbia University have observed that nicotine in the bloodstream of a pregnant rhesus monkey impairs the heart rate, blood pressure, oxygen supply, and acid balance of the unborn. After a two-year study, the scientists suggested that babies born to heavy smokers may be smaller, lighter, and perhaps less healthy. Statistical studies of human mothers indicate that smoking increases the chances of natural abortions and premature births. Statistically the weight of a child born to a smoking mother is significantly less, averaging about six ounces less than the newborn of a nonsmoking mother. If your child had been carried by a heavy smoker and came into this world frail and underweight, would your suspicions lead you to blame the natal mother, especially if she were your sister-in-law? One final question about substitute mothers, which hardly exhausts the questions we should ask: would the anonymous foster mother have a claim to a child whose genetic parents died in an automobile accident, and would that claim outweigh that of close relatives of the genetic parents?

There are a lot of questions we can ask about the morality and human values involved in these possibilities. (There are undoubtedly many more that we have yet to think of and should ask.) Even in the traditional context of a fixed morality where everything about human sexuality is focused on procreation, the ethics of some of the procedures outlined above are far from clear and certainly not easy. For instance, if the whole rhyme and reason behind human sexual intercourse is procreation, is it immoral for a doctor to use his technical skill with inovulation to bypass a blocked oviduct? Is it immoral for him to collect an egg from this barren but fertile married woman, fertilize it *in vitro* with her husband's semen, and implant the resulting embryo in her womb? How should we view the morality of such a solution to the childless marriage? Is it morally and humanly preferable to postnatal adoption? What of the morality of prenatal adoption when this involves an anonymous donor for the egg, fertilization with the husband's semen and implantation in the wife's womb?

Society will soon have to appraise the morality of a substitute mother. How will we react socially to a woman we learn has been a substitute mother? Will we be tempted to gossip and make snide remarks behind her back to the neighbors? What if she is an unwed foster mother for her married sister? And how will we judge the ethics of using a subhuman surrogate mother? Morally and in terms of human values is a subhuman foster mother better, worse, or on about the same par as an artificial womb pregnancy? For thousands of years human beings have accepted the often necessary custom of the wet nurse. Long before the advent of Pablum and Similac wet nurses were viewed as the epitome of Christian charity. Not only were they utterly womanly in their instinct and behavior, they were able to help their neighbor in a way no man could. This was very laudable in the Judaeo-Christian morality as well as universally. In essence the wet nurse simply provides necessary nourishment for a newborn which the natural mother cannot supply. Parallel this situation with that of the surrogate mother. The substitute mother simply provides for a fertilized egg the necessary nourishment and nursery for an unborn which the genetic (natural?) mother cannot supply. Is the surrogate mother then a mere extension of the wet nurse? Does she fall under the same morally acceptable umbrella? Does it really matter whether the foster mother offers her charitable services before or after birth?

One of the more debatable, titillating, and likely least productive questions stirred up by the possibility of surrogate mothers is the applicability of classic concepts of sexual morality such as adultery, marital infidelity, and incest. Can we, for example, raise the accusation of incest when a husband and his sterile wife agree to his or her sister providing an egg to be fertilized by his semen and implanted in his wife's womb? Can the classic moral judgment against adultery be applied to the passionless scientific procedure of *in vitro* fertilization? Or to artificial fertilization, *in vitro* or *in vivo*, when the egg or sperm of an anonymous donor is used?

During the Middle Ages usury was considered by many utterly immoral because inanimate objects like money cannot grow or produce fruit. Yet the mercenary soldier was taken for granted. I wonder how we will react to the mercenary end of the surrogate mother, when and if this becomes a reality of daily life. Perhaps we may be forced to control by legislation the rental fee on natural and artificial wombs. Many surrogate mothers, especially those who are closely related and known to the genetic parents, will perform their charitable services without thought of recompense, out of selfless Christian compassion. On the other hand, if males today are offered a set fee for their contributions to sperm banks, should not the donor of an egg also be reimbursed for her inconvenience? Take this one step further. A childless couple may want to purchase an already fertilized egg. What then? Who reaps the profits? Who sets the fee? Why settle for second best when you are in the market for a fertilized egg? As the successful Emil told a friend in James Costigan's play, *Baby Want a Kiss*, "Celebrity Seed" is the thing of the future. You too can have a child, the product of the sperm and egg of your favorite sports star, movie actor and actress, TV or radio personality, congressman, writer, artist, or scientist. You select the combination, and the wife carries the product. Only remember your budget. In the area of surrogate mothers it is hardly reasonable to expect an anonymous woman to go through the inconvenience, medical examinations, and fees of a nine-month pregnancy without recompense. Nor is it likely that you can hire her for a forty-hour week as the usual pregnancy is a twenty-four-hour-a-day, seven-day-a-week affair. That would mean room and board, at least on a prorated basis, along with reimbursement for fees and the like.

There are other questions, besides those touching the moral values of man, which we should ask from a psychological vantage point, though the moral issue of human values weaves through these also.

Five of thirteen heart transplant patients at Stanford University have become psychotic while three others have shown signs of organic brain damage. Similar cases of mental illness have been reported by psychiatrists observing transplant patients elsewhere. Generally the reaction includes paranoia, refusal to take the medicine which prevents rejection of their new hearts, yanking bandages off, and pulling the intravenous tubes out of their arms. Two severe psychological reactions have even been reported from the *parents* of Houston heart transplant patients. Much of this can be traced to or blamed on the drugs used to suppress the rejection of the new heart, especially prednisone. But how much of it may actually depend on the psychological bent of the patient, exclusive of the pharmaceuticals used? No one knows, yet similar side effects are possible when the artificial womb and surrogate mothers become a reality. No one today with any competence would venture to set out in detail the psychological traumas and reactions the first surrogate mother is likely to face. We can and should try to estimate these, but at the same time realize how tentative and exploratory our guesses are. One bizarre sidelight to the heart transplant phenomenon is the number of people who call, write, or actually appear at the hospital offering to donate their hearts for transplant on the spot. Serious consideration certainly would have to be given to some psychological screening of prospective substitute mothers. Perhaps they would have to be certified and licensed by some agency. Bonding agencies and insurance companies might even get in on the act.

The psychological reaction of the fetus is another point we must consider. How is the fetus likely to respond to a mercenary mother, especially if his nine-month incubator is of the bovine or simian species? Would he recognize the substitution and subconsciously react, biologically, hormonally, or biochemically to the foreign womb? How is his human body and mind likely to accommodate itself to the hormonal differences of nine months in the womb of an animal of another species? Postnatally, how might the child react? Would he look with pride on his nine-month prenatal life grazing in the backyard as being a mark of distinction, sophistication, and aristocracy, or might he come to resent his mother who put him out to pasture while she played gourmet cook and cultured hostess, lover and intellectual companion to his father? Newborn lambs are psychologically and biochemically conditioned or imprinted to respond to their natal mother. How much of this imprinting occurs before birth we do not know, but it is interesting that an orphaned lamb will not nurse with a substitute mother unless fleece from its dead mother is tied around the foster mother's neck. Perhaps we may face the same rejection in a human child who spent nine months in a subhuman foster mother.

Some of the possibilities and questions posed above may sound absurd, grotesque, hideous, and downright repulsive. Perhaps rightly so. Still there is a dimension which is often ignored by journalists, theologians, and laymen alike, namely that the same adjectives were used not too long ago in very similar contexts to describe the public reaction to blood transfusions, kidney and heart transplants, childbirth by anesthesia, and then by a round-about-face natural childbirth and breast feeding. Taking the perspective man's history offers makes one wonder quite seriously whether the emotional reaction to the above possibilities may not, ten or fifteen years from now, fall into the same box of anachronistic and curious mental patterns as our past objections to contraception, operations, transfusions, and transplants. Later on we will return to this subject from a slightly different angle to ask how "natural" or "unnatural" these developments may be in terms of man's sexuality and nature.

However we view the moral and ethical issues raised by surrogate mothers and embryo transplantations, one thing is certain and evident, namely, that we must weigh carefully the pros and the cons *now*, before deciding what to do. Whether we outlaw these developments or legitimize them socially and legally, we will still face the exhausting task of exploring all their angles and implications. That is, if we want to come to a rational decision on the matter. How do the potential dangers and hazards stack up against the advantages to be gained?

Some of the points raised above sound horribly futuristic and unbelievable. But then those same questions may become a reality to be faced within the next decade or two. Forecasting the future is terribly risky, and many a prophet has gone down to an ignominious death of irrelevance when the tide of human affairs shifted. But I think there is something much more basic than surface currents in the topics we are exploring. As I mentioned before, the new technology of human reproduction cuts to the very heart of man's nature and life. It is not something we can ignore if we find it too repulsive to think about. We have no choice. We have to face it sooner or later. We can admit that it is a bit unrealistic and fantastic to talk *today* of meeting a substitute mother at the local department store, or looking out in the backyard to check on the progress of the latest addition to the family circle, or dropping by the hospital to see how junior is doing in his womb of glass and steel. Perchance it is too early to wonder about the advisability and morality of newlyweds superovulating, inseminating, collecting the embryos, and implanting one or two now to start off the family while storing the rest for later use. It is likely too early to wonder whether your wife should super-ovulate and have triplets or quintuplets the first time around so that your family circle is complete in nine short months. But how premature are we? Can you predict, looking back at the lesson of the contraceptive pill and its history?

While we muddle along here trying to cope, emotionally and intellectually, with some of the new developments, research in the medical field and in the biology laboratories continues around the world at a rate one can only describe as exponential. Every year new developments come into view faster and faster; the pace of change is undeniable in its geometric acceleration. So while we discuss certain possibilities as being just around the corner, a decade or two off perhaps, the scientists look at us and say, "How quaint you are, still worried about those *old* problems." Etienne Wolff, one of France's leading experimental embryologists, has highlighted this awareness gap by pondering the eventual emergence of a whole subculture or race of incubating mothers who may very well take over the task of carrying children for the rest of the human species, while the rest of us ponder the morality and implications of the "pill."

In animal husbandry, inovulation may well prove to be man's most priceless advance. In the past fifty years at least twenty species of mammals have disappeared forever from the face of our land. Over four thousand species of mammals are today on the verge of becoming tomorrow's fossils. Man has created his own environment, but in doing so he has become a deadly threat to other animal species. Only a few hundred giant sable antelopes survive today in West Africa. The banded anteater of southwest-ern Australia has become the defenseless prey of dogs, cats, and sheep introduced by man. Less than four thousand tigers are alive in India today and there are likely more orang-outangs in zoos than free in their native Borneo. The Tasmanian wolf, a marsu-pial like the kangaroo, has fled to the remote reaches of the Australian outback to escape the overwhelming competition of wolves brought by man to that southern continent. The Euro-pean bison, cousin of the American buffalo, was down to less than a hundred after the Second World War; today there are about four hundred living in game reserves and zoological parks. The condor, the American bald eagle, the cougar are also among the animal species threatened with extinction.

Sheltering these endangered animals in parks and reserves is often no solution, since many of them refuse to breed when confined or taken out of their native environs. Only recently have some zoologists joined forces with experimental embryologists to explore the possibility of using superovulation and embryo transplants to save these animals. The hope is that by superovulation and *in vitro* fertilization the scientist can obtain enough fertilized eggs to transplant into substitute mothers of a related but more plentiful species. If, for instance, the European bison could be superovulated, a hundred or more eggs fertilized with frozen semen, and the embryo centuplets shipped by way of the Parkes/Hafez "handy thermostated containers," a rabbit incubator, to the United States, they might be transferred to wombs of the much more plentiful American bison. The European bison population could raise by 25 percent at one stroke. Think of the frustrations that could be avoided on both sides of the Iron Curtain if the British and Russians could superovulate and inseminate their two cantankerous pandas.

The applications and advantages of surrogate mothers and inovulation to animal husbandry are almost without limit. Fortunately, ethical questions there are not nearly so complicated as they are in the human sphere. The application of eugenics to improve the quality of herds, their milk- or beef-producing ability, are taken for granted. The use of these techniques with animals is not overshadowed by the question of abortion even when the studies to discover the causes of prenatal mortality and miscarriage involve aborting hundreds or thousands of fetuses.

Before we leave this field for other developments we should say something about the possible transplantation of the reproductive glands themselves, the ovary and testes.

The human ovary is an incredible organ. Sheltered in this structure, the size of a large almond, are half a million eggs, of which only five hundred will be released during a woman's fertile period of life. Of these five hundred many fewer than a dozen are usually fertilized though the reported record for fertility in a single woman is fifty children. (That was without

multiple births being induced hormonally!) At the same time
the ovary is producing germ cells, it is an essential component in
the hormonal or endocrine system of the woman, linking, responding to, and triggering a variety of biochemical messengers
in the pituitary and adrenal glands, the uterus and breasts, which
in turn influence the whole development of the body and its
psyche.

Briefly let me trace some of the functions of the human ovary.
Once activated by the two gonadotropic hormones, FSH and LH,
the ovary enters into a process centering on monthly cycles. The
follicle-stimulating hormone, FSH, triggers the development of
one or more follicles in the ovary. The nurse cells of the follicles
in turn produce an estrogenic hormone, estradiol, which brings
about changes in the epithelial lining of the vagina. Under its
influence cells in the mucosal lining of the womb also proliferate
as the surface becomes "spongy" in preparation for a possible
pregnancy. When the egg breaks out of the mature follicle, another team of hormones goes into action. The luteinizing
hormone, LH, from the anterior lobe of the pituitary (hypoph-
ysis), persuades the cells of the ruptured follicle to change
their chemistry and become tiny factories for the hormones for
maintaining pregnancy should this occur. As the egg is passing
down the fallopian tube, the hormone progesterone from the
ruptured follicle brings the spongy tissue development of the
womb to a high point, ready for implantation if fertilization occurs. When implantation occurs, the mucosal lining of the womb
and later the placenta send hormonal messengers back to the
corpus luteum as the ruptured ovarian follicle is now known
and to the pituitary gland just under the brain. The *corpus
luteum* will suppress development of any new follicles if pregnancy has begun. If not, it degenerates and a new cycle brings
the maturation of a new follicle. With pregnancy the pituitary
gland releases into the bloodstream yet another secretion, the
lactogenic hormone, which prepares the mammary glands for
milk production and urges the *corpus luteum* in the ovary to
increase and maintain its secretion of progesterone for the uterus.
This is an extremely sketchy diagram of the ovary as a hormone-
secreting gland. It does not mention all the question marks, of
which there are many, nor the disputed links and feedback

mechanisms. But it is detailed enough to allow some appreciation of the ovary in human physiology and some indication of the disastrous effects its absence or malfunction can have on a woman.

Whole ovaries have been maintained in tissue culture media for several weeks by Dr. Theodore Fainstat of the Strangeways Laboratory. In 1966 Dr. Fainstat found that at least temporary culture could be achieved if the oxygen supply was reduced to correspond with the stage of development the ovary is in when removed from a patient. Other scientists have tried to culture small clusters of cells or even single ovarian cells, particularly those which make up the ovigerous cords that give rise to the primary germ cells. Thus far they have managed to induce these cells only to produce progesterone. Nevertheless, ovary banks are feasible and a serious possibility in the near future.

Several attempts have been made to transplant pieces of ovarian tissue. Dr. S. H. Sturgis, professor of gynecology at Harvard University, has studied possible ways to help women with abnormal or nonfunctioning ovaries. Working first with rats and monkeys, and then with women, Sturgis tried embedding tiny pieces of ovarian tissue in a permeable material before transplanting it into the malfunctioning ovary. His hope was that the healthy transplanted tissue would avoid rejection, become hormonally active and coordinate the malfunctioning gland. There was some evidence of ovarian stimulation, but not enough to prompt Sturgis to continue the experiments.

More recently, however, Dr. Hector A. Castellanos, a Temple University professor, reported some success with the Sturgis technique. Having worked under Sturgis at Harvard, Castellanos used healthy ovarian tissue taken from women undergoing sterilization for his transplants. The fresh ovarian tissue was first minced for mixing with corneas that had been rejected as unfit for corneal transplants. His hope was to relieve women suffering from symptoms comparable to menopause because of nonfunctioning ovaries. As of late 1968 Castellanos had succeeded with twelve such operations. Eventually he hopes these and other women in the same situation may be able to bear children of their own. One of the problems Castellanos is working on is a remedy, through ovarian transplantation, for the sizable number of women whose ovaries are manufacturing the male hormone androgen rather than the female estrogen. By removing the abnormal ovarian tissue and replacing it with healthy normal tissue, the Temple scientist hopes to make ovulation and childbearing a possibility for these women. Transplantation of whole ovaries is already a reality in dogs where defective and atrophied canine ovaries have been rejuvenated by being transplanted to healthy young females. As with the fetal transplants, ovarian transplants appear to succeed because they are basically embryonic tissue and less sensitive to the rejection phenomenon.

So we face yet another problem and some revolutionary questions for ethics come to the surface of human consciousness. Rostand has asked about the case of a barren woman who receives a healthy ovary donated by another woman and subsequently conceives a child. Is the child actually hers, just because she carried it? Or is the true mother the woman who donated the ovary and whose genes the child carries? Or are both women mothers in the same or in different senses of the term? We can carry this one step further, as Glanville Williams has suggested, to ask about motherhood in a case identical to the one above except that the egg was fertilized by sperm not from the woman's husband but by semen from an anonymous donor. And if we were to use frozen semen from a man deceased several years, what then?

On the other hand, what happens if scientists are able to transplant testes from one man to another? Try to figure out with some semblance of logic the following case. A barren woman, citizen of Russia, receives an ovarian transplant from a Negro citizen of Nigeria. Married to a sterile man, a native of the Australian bush country, she is artificially inseminated with frozen semen from an Eskimo who died eleven years ago. But the Russian woman has difficulty in carrying the child so she arranges for an American Indian woman to serve as a substitute mother. Nine months later the child is born. Puzzle out, if you will—and if you can—the racial and national constitution of the offspring, its citizenship, and its two (?) parents!

Legally, we have no guideline whatsoever with which to judge the above case. A British legal review, *The Solicitors Journal*, has offered the only bit of counsel, and a very small bit at that, we have on this subject. It reported that a well-known British gynecologist was ready to try an ovarian transplant on a woman until the Medical Defense Union informed her that any offspring that might result would be considered illegitimate. For the British, at least, ovarian transplants could have the same legal effect as artificial insemination bears in some states. But are the two situations so parallel in terms of logic and common sense?

How are the man and woman in the street reacting to all these developments? *Basically, they are not reacting,* for the simple reason that very few of them are aware that such developments are possible, let alone already or soon to be applied to human reproduction. Ten years ago I wrote an article on "The Challenge of Utopian Biology" for a small Catholic magazine. Though I discussed a number of radical questions about human reproduction and its future, the readers seemed to ignore the whole issue. Jean Rostand, Julian Huxley, and others wrote about the biological revolution, but the time was not ripe for such an idea to take hold. In 1965 *Life* magazine explored "The Control of Life" in a provocative four-part series authored by Albert Rosenfeld. But again, while there was some response from the general public, some curiosity, and a very modest amount of serious discussion, the impending challenge of "utopian motherhood" was generally ignored. Over the long range, I think, the rapid acceptance of the contraceptive pills with their radical impact on family life and male-female relationships and the sudden emergence of heart transplants has now brought the subject to germination in the public mind. The idea of man's accelerating control over his own reproductive capacities has come of age.

Even so, the general public's ignorance of new trends in human reproduction is more than evident in a recent nationwide survey involving a representative cross-section sampling of both urban and rural American men and women. Only 3 percent of the 1,600 adults surveyed were aware that a doctor can use a donor's semen to inseminate a woman without intercourse occurring. In reality this ignorance of artificial insemination is not surprising, despite the existence of over 150,000 Americans who owe their being to this practice. Parents who have resorted to artificial insemination are not likely to advertise that fact beyond their immediate family circle, and perhaps not even that far. To compensate for this lack of basic information, Louis Harris and Associates prepared a lengthy questionnaire giving all the necessary background information as preface for each specific question. While the only procedure feasible under the circumstances, this interview technique does allow considerable latitude for hidden moral persuasions and bias. The wording of the prefacing statement can easily sway the respondent one way or the other. This cautionary note is particularly appropriate here since the Harris survey, undertaken for *Life* magazine and published in its June 13, 1969, issue, is the *only* evidence we have of the public reaction to these new trends in human reproduction.

The overall results of the survey are in a way surprising. We are dealing with some quite radical changes in human life, as the discussion thus far makes very evident. Furthermore, these developments have been for the most part unknown to the general public. A natural reaction to any discussion of them would involve strong reservations and might likely be even predominately negative. The first time one hears a radical change proposed for one's own personal life the immediate and natural response is seldom open acceptance or even cautious embrace. Yet the Harris survey revealed "a remarkable readiness to accept revolutionary values and modes of behavior" indicated by these new modes of human reproduction. While two out of three people rejected the new developments overall, one-third of the respondents did approve of them. Extrapolated over the total American population this would mean that nearly one billion Americans basically could live with the new forms of parenthood.

As you might expect the greatest degree of acceptance was expressed by the younger generation and by the better educated segments. The texts and lab manuals of the Biological Sciences Curriculum Study (BSCS) program, among the most popular biology texts in American high schools, have been introducing young students to the very latest in experimental research including the artificial insemination of frogs since its pilot program began in the summer of 1960. Having heard about artificial insemination during their formative years would tend to accustom today's young adults to the possibility that it might easily become a part of human culture and even touch them personally later on as married people. Their parents, on the other hand, have learned of this development much later in life, long after their thought patterns on human sexuality and family mores have been pretty much set. This difference was brought out in several ways by the Harris *Life* poll. Fifty-nine percent of those with only an eighth-grade education rejected any tampering with human reproduction at all. Black Americans were generally more traditional and moralistic about the changes than were white Americans, except for *in vitro* babies where the Negro evidenced a strong interest and curiosity. In cases where the wife is infertile, 57 percent of the women under thirty approved of egg implantation. Only half as many, 30 percent, of those with an elementary education agreed.

Throughout the survey Harris found two dominant and crucial factors influencing the decisions. These two factors, in a way, reveal the *ambivalence* of the present situation and of the future of human sexuality. The trends we sketched at the close of chapter one are encapsulated in these two factors. As Harris pointed out, there is a great fear of infertility among Americans and particularly a fear of infertility as a threat to the stability and permanence of marriages. This fear instinctively inclines people to accept any technique that will eliminate or reduce infertility. One-third of the families represented had experienced trouble in childbearing and one in five of the women had had miscarriages. At the same time, most Americans see in these new methods of reproduction a real threat to marriage. Sixty percent of those polled feel that the new approaches to childbirth constitute "the beginning of the end of family life" and the disappearance of procreation through love. On this second factor, 50 percent of the adults reported knowing an unfaithful husband or wife and 21 percent admitted that it was a person close to them, in most cases within their own families. What a classic dilemma! Infertility is a serious threat to a stable marriage, yet the means we have to remedy infertility may themselves prove more disabling to the traditional marriage than the original problem. Are we then faced with a cure that is worse than the disease?

Children and the couple marriage, the monogamous family, clearly appear in the Harris *Life* poll as *the* central motivation of the majority of Americans today. The more radically a technique departs from this golden goal, the less acceptable it is. The

RESULTS OF THE *LIFE* POLL

conducted by Louis Harris and Associates, Inc.
(published June 1969)

ACCEPTANCE OF NEW
METHODS OF REPRODUCTION

	Men	Women
Approve hormone treatments	60%	70%
Approve psychiatric help	63	65
Agree to insemination with husband's sperm	49	62
Approve insemination with anonymous donor sperm	24	28
Approve egg implantation	32	39
Approve *in vitro* babies	25	25
Anonymous donor justified if husband is infertile	33	37
Egg implant justified if wife is infertile	34	43
In vitro babies justified if wife might die or be crippled from childbirth	30	35
Opposed to all three new methods	42	39
Morally wrong to have babies through new methods	45	40
Would feel love toward *in vitro* baby with own sperm and egg	47	53
Would feel love toward egg implant child of own	37	48
Would feel love toward *in vitro* baby not of own sperm and egg	30	38
In vitro child would feel love for family	55	61
Know unfaithful husband or wife	51	50
Unfaithful person "close to me"	20	22
Incidence of miscarriage or trouble conceiving in own family	27	41

more a new method appears as a threat to the family tradition, the more widely it is to be rejected. Thus, 65 percent of the Americans surveyed approved of hormone treatments or psychiatric help to remedy an infertile marriage. Forty-nine percent of the men and 62 percent of the women agreed to insemination with the husband's sperm. Artificial insemination with a donor's sperm was accepted by only 19 percent. When the situation was

CONTROVERSY AND CONFLICT
OVER THE NEW REPRODUCTIVE METHODS

(The percentage of agreement and disagreement is given in parentheses)

MORALITY
They are not morally wrong (47% agree, 42% disagree)
   BUT They are against God's will (53% agree, 34% disagree)

PARENTHOOD
No family need be childless anymore (51% to 35%)
   BUT It would mean the end of babies born through love (71% to 21%)
   And it will reduce love-making to a physical act (54% to 33%)

FAMILY LOVE
A child born through new methods would feel the same love
                                        for its family (58% to 18%)
   BUT It is more natural for parents to love a normally
                                        conceived baby (51% to 37%)

KIND OF BABY PRODUCED
The new methods are genetically safe (59% to 18%)
   BUT It is wrong to breed children for special
                                        characteristics (67% to 20%)

New methods can be used to avoid birth defects (58% to 26%)
   BUT It is wrong to produce superior children from superior
                                        sperm and eggs (57% to 21%)

ADVANTAGES
More convenient (50% to 40%)
   BUT It is wrong for women to avoid problems of
                                        pregnancy (80% to 11%)

POPULATION CONTROL
Overpopulation can be avoided (46% to 39%)
   BUT Regulation of who can have babies is wrong (60% to 23%)

STRAIN ON FAMILY
Husband and wife would love each other no less (52% to 26%)
   BUT Men will feel emasculated (76% to 17%)

It will not lead to more infidelity (45% to 35%)
   BUT It will encourage promiscuity (51% to 29%)

qualified by pointing out that this was the only way a childless couple could have a baby of their own, 35 percent approved. General acceptance of artificial insemination with the husband supplying the germ cells was 55 percent; with a donor supplying the semen, only 26 percent. Men have funny egos, as one of the participating women remarked, and artificial insemination can easily pose a psychological threat to a man's masculinity. The use of a donated egg is not viewed with nearly the same fear since the woman's ego and femininity are supported more by the fact of her nine-month pregnancy than by the knowledge that her egg is responsible for that child. In this contest of masculine insecurity and fears for fidelity in marriage, more than 90 percent of those polled turned thumbs down on artificial insemination when the husband is in prison or away on military duty.

Interestingly enough one of the most radical departures from traditional reproduction found considerable, if still minority, acceptance: the artificial womb. Twenty-five percent of the men and women found this prospect acceptable and 23 percent said that they might use it themselves. Again circumstances changed the response. If the husband or wife were infertile, 27 percent would approve of an *in vitro* pregnancy; if childbirth would pose a danger to either mother or child, one-third of those polled agreed to its use. Particularly on this question, motivations were carefully weighed. If a woman "just wants to skip pregnancy and have a baby too," more than 90 percent of the men and women would not approve. A lot of skepticism was expressed about career women and the relationship between motherhood and a career outside the home. Only 5 percent thought a woman would be justified in having an *in vitro* pregnancy so she could continue working and still have a family.

Another aspect of the basic concern expressed in the poll, both infertility and its remedies as threats to the family, came across in the repeated rejection of new developments which threatened to change the mother's traditional role. But in this *women seem much more willing to change than men.* The analysis of the Harris poll presented by Albert Rosenfeld passed over this point which to my way of thinking is *as important as it is evident* in the results tabulated from the poll. *In every question the women were significantly more open to change than were the men. Since women are the ones most directly concerned with reproduction, their views are likely to carry much more weight in terms of our future direction than are the views of men.*

The biological developments surveyed thus far pose few serious problems as long as they are used with animals outside the human circle. Experimentation for experimentation's sake there is fine and few people object when a serious scientist pursues the quest for new knowledge by artificially inseminating a cow or using a rabbit as incubator for a hundred lambs. Economics often settle any moral questions that arise with animal experimentation. But scientific and medical research cannot be fenced off or isolated from the world of man. When some new technique proves useful with animals the question automatically comes up as to whether we might also use it for the good of man and mankind. Inevitably, such a question stirs up a hornet's nest of controversy and conflicting opinions even within the mind of one individual.

Biologically or medically there is little quarrel with the new trends in reproduction, but when applied to man the moral, religious, and human implications set our minds spinning with confusion and conflict and controversy. The sixteen hundred Americans surveyed by Harris offer a beautiful miniature of the tensions and conflicts all of us experience in one way or other in this present discussion. Many of them found the new methods morally acceptable, though against God's will. Many were convinced that the childless family was no longer inevitable, while the new methods which remedy the childless marriage take love out of procreation and reduce human sexual intercourse to a mere physical act. Many believe that a child born through the new methods will feel a normal love for its family, but that parents of such a child would find it more natural to love a child conceived in the normal way. Fifty percent thought the new methods were more convenient but 80 percent were convinced that it is wrong for women to avoid the problems of pregnancy. Overpopulation should be avoided but the regulation of who can have babies or how many you can have was soundly rejected. While the new techniques might make men feel emasculated, the same people were convinced that the husband and wife would still love each other as much as before. And then the conflict to top them all: many thought that the new methods of human reproduction would encourage promiscuity while at the same time not leading to more infidelity. Sic!

So much for the only poll of public attitudes on the new embryology. It is difficult to estimate how valid this survey is in representing the views of Americans, or by extrapolation the opinions of Europeans or other peoples. Even on the American scene, public opinion is changing very rapidly as more and more popular journals and women's magazines carry articles outlining the new developments. It is far too early and we have far too little evidence to essay any prediction of how Americans or other peoples will react in the days to come.

Up to this point we have taken a rather serious approach to some very serious possibilities in human reproduction. Touching every human deep in the core of his being, these developments raise razor-sharp dilemmas in the midst of our most intimate social relationships and structures. They touch men and women in their most personal sexual nature. Hence a sober and calm examination of the issues at stake is essential and indispensable. The questions are serious and we must treat them that way. Nevertheless, there is something about human sexuality which makes the completely serious discussion often overwhelming for many people. With that in mind, we might lighten our discussion for a moment with a brief excursion into three recent experiments which contain some humorous overtones and possibilities.

Most people within the range of the news media are conscious of an exploding human population. But some experimental embryologists have apparently been so engrossed in their ivory tower laboratories that they have missed this foreboding news. As a consequence their research projects, on occasion, tend to come up with rather impractical solutions to nonproblems. Take, for example, the efforts of Drs. Lutz and Wolff to create artificial quintuplets from a single fertilized egg. The duck egg provided these scientists with a perfect guinea pig since all birds develop in fairly large eggs where the embryo proper is easily located on top of the yolk in a large obvious embryonic disk. Wolff and Lutz wielded their scalpels so skillfully some years back that they were able to slice up the primitive streaks of some duck eggs into four or five pieces. The primitive streak, as you may recall, develops into the central nervous system, brain, and spinal cord. It is also one of the primary "organizers" of embryonic development, inducing other undifferentiated tissues and cells to form specific organs and structures. By the time the two scientists had finished they had neatly turned single ducklings into quintuplets.

Lutz and Wolff engaged in the old biological game of multiplication by division. The reverse of this new biological mathematics, a case of subtraction by addition, occurred recently at the Institute for Cancer Research in Philadelphia. Dr. Beatrice Mintz has been working there for some five years with "muddled mice." Despite my playful treatment of this experiment, I must emphasize that it does have some very important and serious scientific applications in cancer research, genetics, and immunology. My vagrant interpretation of Mintz's work stems more from an aftermath of a bizarre combination of gourmet and gourmand delights late one summer evening. If fifty years ago an Oregon physician could accuse Dr. Addison Davis Hard of digestive upset in his description of Dr. Pancoast's venture in ethereal copulation, then I am sure he would find much more justification for this charge in my report of Dr. Mintz's work.

Dr. Mintz began with two distinct purebred strains of mice, an albino and a black strain. A pair of each strain were mated with their own kind, their fertilized eggs collected from the fallopian tubes, and dropped at the eight-cell stage into a tissue culture medium containing some fetal calf serum and other "spices." Pronase, an enzyme in the medium, gently dissolves the *zona pellucida* which encapsulates the zygotes, after which they are washed and returned to culture dishes. Each dish then contains side by side two embryos, one from each genetic strain mating. Within ten minutes, the paired zygotes adhere to each other. Placed in an incubator, each combination mass of cells slows down its development so that after twenty-four hours a normal thirty-two- to sixty-four-cell blastocyst has developed. Dr. Mintz then transfers the embryos to the womb of a pseudopregnant mouse. There about one-third of them develop into viable mice.

Christened by scientists with the esoteric name of allophenics, these mice are unique in much more than name. Each allophenic mouse has *four* parents, two black and two albino parents. Each allophenic mouse has two different sets of genes and chromosomes. Not that each cell in its body has two sets, but there are two distinct strains of cells in the mouse, each strain (clone) of cells with its own normal set of genes from the original mating.

One wonders why all the fuss about the population explosion and unsatisfactory contraceptives when we have a solution to the whole mess right at hand. All the scientists have to do is arrange for women who have just conceived to drop by the local hospital within say twenty-four hours. The resident physician collects her tiny offspring in a syringe and holds it in cold storage until a mate comes along of the same sex. (Dr. Mintz has had some interesting results with intersexes, the combination of male and female embryos.) A culture dish, a little fetal calf serum, a dash of pronase, and presto, two humans are now one human. The two mothers can then flip coins or draw straws to see who carries their child the first four and a half months and who takes over for the second half. Inovulation, a later transfer, then birth. Besides cutting the birth rate in half and thereby solving our population explosion, this technique has the added advantage of practically eliminating the orphanage. Each child would have *four* true natal and genetic parents! His chances of being orphaned are greatly reduced.

The second major (hypothetical) advantage of this experiment arises from the phenotypic character of the mosaic allophenic mice. The allophenic mouse comes originally from two distinct embryos with different genes for hair color and other traits. Its two types of cells do not merge but rather coexist side by side in the combined embryo. This means that the newborn allophenic mouse ends up a muddled mosaic in black and white. Some aesthetically interesting color combinations have occurred since the first mosaic mouse was born on New Year's Day in 1965: vertical and horizontal strips, round and irregular spots, different colors. Which makes me wonder about a possible application in the theater. If the nudity rash of the 1969–1970 Broadway theater continues as something more than a fad, it will have to develop some variety. Aesthetically there is only so much one can do in terms of natural color combinations in plays like *Hair*, *Oh Calcutta*, *Che*, and their offspring. The bare human torso can be mighty boring after long exposure without variety. We might suggest then that the Actors' Guild enter into an alliance with some experimental embryologists to produce allophenic actors. Think of the variety this could lead to! Red and yellow stripes, vertical or oblique; red and black speckles; and so through the rainbow.

Finally, there is the case of the "green mice" which came out of the Department of Therapeutic Research at the University of Pennsylvania School of Medicine. Dr. Margit Nass has cultured fibroblasts or L cells from a mouse with the chloroplasts of spinach and African violets. The chloroplasts contain their own chromosome or DNA coding so that they can reproduce independently of the nuclei of cells. When suspended in a culture with the mouse cells, these tiny organelles were ingested by the cells, thus producing "green mouse cells." The chloroplasts are responsible for photosynthesis in green plants and produce basic carbohydrates (sugars) from carbon dioxide, water, and the energy of sunlight. These "green mouse cells" should theoretically be able to manufacture their own food. If this kind of hybridizing were successful with whole mice, instead of mere mouse cells for five generations, we might end up with mice capable of manufacturing their own food from mere water and carbon dioxide, only to get enough sunlight to the chlorophyl we might have to use a hairless breed of mice. With humans clothing might present a similar complication, though the recent see-through fashions might resolve that problem, allowing enough sunlight to filter in to the chloroplasts for them to carry on their work. After some centuries of evolution and natural selection, we might end up with green mice (and green humans) lacking any digestive tract, stomach or intestine. No more peptic ulcers, upset stomachs, Alka-Seltzer, Tums, or Rolaids! In the human arena this technique could complement the Broadway variety mentioned above with allophenic mice and men. Green, red, yellow, brown, black, and white—wide enough variety for the creative producer and playwriter. Fortunately, we are far from that day, however, for the green chloroplasts have not survived with their photosynthetic powers intact in the mouse cells for more than five cell generations, about five days.

But enough of this fantasy. There are still several serious topics we must deal with in our exploration of new trends in human reproduction. We have yet to say anything about the problems

and implications posed by new techniques in monitoring the life
of the prenatal and the possibilities that man may soon completely bypass sexuality and reproduce by asexual cloning. Also
we have yet to tackle the more basic question of what all these
techniques suggest for our understanding of man's sexual nature
and its future. So let us return to the serious vein and move
onward.

FOUR

SEX TAKES
A VACATION
*The Case
for Virgin
Conceptions*

〰〰

In the year 1787 the village of Spandau, Germany, was rocked
by scandal. The assistant headmaster of the local school had just
published a disgraceful volume with the brazen title of *Das
entdeckte Geheimnifs der Natur im Bau und in der Befrachtung
der Blumen. The Newly Revealed Mystery of Nature in the
Structure and Fertilization of Flowers* upset all European society
for in this book the author, Christian Konrad Sprengel, destroyed
the lovely and almost universal illusion that the flowers belong
to the chaste sexless world of the Garden of Eden. Sprengel introduced his readers to the intimate life of the floral world, comparing the different structures of a blossom with the sexual organs
of animals and describing the role of bees, insects, and the wind
in the sexual intercourse of plants. Commonplace as this knowledge is today, it was in 1787 a shocking scandal. Sprengel was
dismissed from his teaching position and his book withdrawn
from public circulation. Had it not been for several broad-
minded citizens who hired him as a private tutor for their chil-

dren, the man who discovered sexual reproduction in the plant kingdom would have been reduced to begging for a livelihood. The pure lily, symbol of all that is chaste and sexless, was sullied by the disclosure of its sexual habits, even though its common image remained untainted.

I recall a luncheon discussion a few years back when I was visiting a priest friend in Detroit. The pastor listened with growing rage as my friend and I spoke of some of the "new theology" then being introduced into the seminaries. Some European Catholic theologians had suggested that the Virgin Mary's virginity might have a much deeper meaning than mere biological integrity. Some of these scholars were then suggesting that the authentic biblical meaning of virginity was to be found in the spiritual or moral sense of total openness to all creatures and the Creator in love, rather than in a biological reduction to physical integrity and the absence of sexual intercourse to which most Christians, influenced by a platonic dualism of matter-evil/spirit-good, have reduced it. The pastor's rage finally broke its bounds, and with an evident deliberate effort for control he asked my friend, his assistant in the parish, to "please take a hammer over to the church after we finish eating and knock that lily out of St. Joseph's hand."

But it was not only the lily whose sex life disturbed Western civilized man. For some unexplained reason Western man has found the whole area of sexuality a real embarrassment, practically since the beginning of recorded history. Nearly two thousand years ago, following the custom of both naturalists and philosophers, the Roman historian Pliny divided the animal kingdom in two: the moral animals, those good creatures who shunned as much as possible any sexual activity even in the matter of reproducing their kind, and the immoral beasts. For Pliny the elephant was a true model of moral rectitude: "Out of shame elephants copulate only in hidden places. . . . Afterwards they bathe in a river. Nor is there any adultery among them, nor cruel battles for the females."

Medieval writers ranked the elephant right at the top of the moral creatures of our earth. Albert the Great, whom we had occasion to mention earlier, stated that the elephant was devoid of any sexual instinct, conceiving and giving birth in all the supposed paradisiac innocence of an animal Eden. Two centuries later Konrad von Megenberg, one of the first popular zoologists, contrasted the quasi-angelic habits of the elephant who copulate only for the sake of procreation and then do not touch the female for three years with the frivolous loose morals of those beasts which "live for their lust without divine worship." Even as the age of the Renaissance closed a little over two hundred years ago, another popular naturalist, Konrad Gesner, praised the virtues of the pachyderms: "Chaste and pure they lead their lives, giving way to no luxurie, offending never against their wedded state, and are of speciall shamefulness, for they engage not in such thinges unless before they have enclosed them[selves] in thicke shrubberie and dense undergrowthe."

Would that man were as chaste and pure as the elephant!

Ah, but he was in the beginning, at least according to some theologians. For in the Garden of Eden, God had decreed that man should reproduce his kind without benefit of sexual intercourse. In fact, according to St. Gregory of Nyssa, St. John Chrysostom, and that most learned celibate misogynist, St. Jerome, man would have reproduced by some sort of angelic division of the body, by a chaste embrace, or perhaps at most by a chaste kiss. This was the will of God, but God in his wisdom foresaw the proud rebellion of man and thoughtfully equipped both men and women with the reproductive organs of the animals they would become after their Fall.

St. Augustine of Hippo summed up in the third century of the Christian Era the thinking of many Westerners both before and after his time when he spoke of the ideal human situation "as it should be" and as it would have been in the Garden of Eden had not man fallen victim of his own pride. In Book XLV of *The City of God* Augustine wrote that, had we remained in Eden:

Those [sexual] members [of our bodies] would be moved like the rest [of our organs] by the command of the will, and the husband would be mingled with the loins of his wife without the seductive stimulus of passion, with calmness of mind and with no corruption of the body's innocence. . . . Because the wild heat of passion would not activate those parts of the body, but as would be proper, a voluntary control would use them. Thus it would then have been possible to inject the semen into the womb through the female genitalia as innocently as the menstrual flow is now ejected.

One historian of religion, D. S. Bailey, called the whole ideal of sexual love proposed by Augustine "a grotesque and somewhat repulsive abstraction, not to say a physiological absurdity," yet the theory had endured hundreds of mutations to spring up time and time again in new forms.

Augustine gave his theory an interesting twist when he tried to explain the origin of Eve in the Garden of Eden. In Book VI of *De Genesi ad litteram* he repeatedly calls attention to his quasi-evolutionary explanation of creation, namely that God created everything virtually or in potentiality in the beginning and placed in nature seminal causes or "seeds" which would develop only in their due time and proper sequence. Eve and the whole feminine sex was contained in potentiality in the causes created by God in the beginning, just as was Adam and the whole masculine sex. Working from this explanation of creation, Augustine was quite puzzled then by all the attention the authors of Genesis paid to the origins of woman. Ernest Messenger, a Roman Catholic pioneer in reconciling evolution with theological tradition in the 1930's, has pointed out that in the context of his remarks in Book IX of *De Genesi ad litteram*, Augustine would not have been surprised if Eve had issued from the body of an animal of a species other than Adam, "as we know innumerable kinds of worms or reptiles to arise." But what really puzzled Augustine, apparently, was the idea that Eve came from the body of Adam, an animal differing only in sex, and then *without sexual intercourse*, "for we know of no flesh born from the flesh of any living creature which is so alike to it that it differs only in sex." But the fact that we cannot find in nature a parallel example for this strange formation of Eve did not disturb Augustine for long.

> We seek in the things of this creation a similitude of the woman made from the side of a man, and we cannot find one, simply because we only know how men operate on this earth. We do not know how the angels work in this world by a kind of agriculture.

Augustine went on to explain this remark in some detail suggesting that if supernatural powers could bring the seminal causes of Mary's virginal womb to fruit in Christ, then angelic powers could logically bring similar seminal causes of a woman, Eve, to fruit from the body of a male. ". . . in the former case the Lord was to be born of a servant, and in the latter, a servant was to be formed from a slave."

Much to Augustine's credit we should note that he wound up his discussion of Eve's origins with the modest admission that "I will say what I am inclined to think, without being so bold as to affirm anything."

In the Garden of Eden then, according to Augustine and countless scholars and theologians before and since, Adam outstripped the chaste elephant in virtue, reproducing Eve asexually in some way or other.

In more recent times the theme of human reproduction without the benefit of sexual intercourse has turned up in explanations offered by some theologians to accommodate a historical traditional interpretation of the origins of Eve with the scientific theory of man's revolutionary origins. This was a major concern of theologians in the first half of this century, and men like Dr. Ernest Messenger devoted much study and writing to the subject. At that time, for Catholics and many in the Protestant tradition, the biblical explanation of "the formation of the first woman from the body of the first man" was a major concern if one wanted to integrate his traditional theology with an evolutionary science of man. The Roman Biblical Commission has spoken bluntly to the issue in 1909 including the formation of Eve from the body of Adam as one of the historical items of Genesis which cannot be denied. Today that fallible decree and the interpretation it defended is very much a debate of the past as biblical scholars and theologians of most Christian traditions are no longer concerned with the temporal origins of mankind through the divine creation of a single unique couple. Newer theological explanations of man's origins have been worked out which preserve the original, unchangeable religious message of Genesis without getting bogged down in the question of Adam's rib and the like.*

* A summary of these views may be found in *Evolving World; Converging Man* by Robert T. Francoeur (New York: Holt, Rinehart and Winston, 1970).

It may prove helpful to our exploration of human sexuality and the new modes of reproduction to recall here at least one of the reincarnations of Augustine's theory of a chaste and virginal ideal human race. It appears in a 1931 classic work by Dr. Ernest Messenger, a very controversial and pioneering volume, entitled *Evolution and Theology*. Messenger reached deep into the writings of the Fathers and Doctors of the Christian Church to find there a wealth of interpretations and theological views which could give a somewhat traditional support for newer explanations of dogma more in keeping with the advances of science in the twentieth century. Following the suggestion of Augustine on asexual human reproduction, Messenger ended by reviving the idealistic picture of an asexual Eden. I recall this example from the writings of Messenger not because it indicates that he was just as inhumane and unbalanced as Augustine in his appreciation of human sexuality, its past and/or future—which he definitely was not—but because it will lead us through some interesting pathways toward a better appreciation of that subject.

> Let us now consider how, in the light of the theological principles just expounded, we can explain the origin of Eve, if . . . we accept the historical reality of the "formation primae mulieris ex primo homine."
>
> The progress of the biological sciences has taught us that it is not only the sexual elements (eggs and sperm) which contain the virtuality of the species and of the race to which they belong. The numerous instances of a-sexual generation, in some cases spontaneous and happening in the ordinary course of nature, in other cases brought about by special circumstances due either to chance or to human intervention, lead us to conclude that, in principle, *every cell in an ordinary organism contains, at least radically, the virtuality of the species and the race.*

Messenger went on to cite several examples of parthenogenesis, asexual reproduction, and regeneration in both the plant and animal kingdoms. His whole point was that, in 1931, the scientific knowledge about cases of natural and artificially induced parthenogenesis offered the theologian of the day a convenient and acceptable explanation of the origin of Eve from the body of Adam, in a very traditional though updated Augustinian approach.

> To make this still clearer, we can imitate St. Augustine and have recourse to the analogy between a-sexual generation and parthenogenesis, but of course in the light of modern science. There exist in nature many cases of normal parthenogenesis. Further, we have known now for some years that ova which normally only develop when fecundated, can nevertheless develop parthenogenetically under the influence of certain chemical, physical, or even mechanical stimuli. Now that parthenogenesis has been realized experimentally with the batrachians [frogs and toads], we can no longer doubt that the virtuality of the species exists whole and entire in the unfecundated ova of all the vertebrates. Will man ever succeed in provoking parthenogenesis and bring it to a successful conclusion in the mammifers? That may well be doubted, in view of the tremendous experimental difficulties. But it is quite likely that agents more subtle than man, such as the angels or devils, could bring about the circumstances required for this end, if it pleased God to ordain that the former should so act, or to permit it to the latter. And who can say that devils have never brought it about in the human species? In any case it cannot be denied that God has the power to bring about circumstances from which [even human] parthenogenesis would naturally follow.

Dr. Messenger's explanation of Adam and Eve in the Garden of Eden and the origins of woman undoubtedly strikes us today as very quaint, odd, and at times full of unintended humor. It is hard for us to realize how serious a problem the origin of Eve was for scholars only a short forty years ago. I almost hesitate to quote Messenger on this point or to use him as an example. In the end I only do so because his reputation as a scholar of his time is still recognized by theologians and scientists alike. (His mentor, the Canon de Dorlodot, was the only official representative of the Roman Catholic Church at the fiftieth anniversary of the publication of Darwin's *Origin of Species*.) The solution proposed by Messenger to the question of sexuality in Eden was offered in a very scholarly tome. That it appears odd and even humorous to us today is only evidence of how far a few decades have taken us in our pursuit of a deeper understanding of our nature and origins.

In any event, I trust these few comments of a historical dimension will provide a background for the scientific aspects of this chapter and their moral, human implications.

Let us return for a few moments to the subject of flowers. A hundred years before Herr Sprengel discovered the sex life of the lily, the Dutch had become fascinated by some insects that seemed to be everywhere in their rose gardens. Chubby but tiny lace-winged green aphids, plant lice, were avidly studied by amateur as well as professional scientists everywhere in the Lowlands. Leeuwenhoek and Swammerdam devoted considerable attention to this little creature, noting that ants tend them very much like cows and milk them regularly while ladybugs serve to keep the miniature herd populations within bounds. The most bewildering fact uncovered by these students of nature came to light in the summer when the roses were infested only by female aphids. No males could be found anywhere, yet the little virgins busily propagated their kind without assistance. From spring to late summer the virgins produced only female offspring. Then as the cool shorter days of autumn arrived males appeared among the progeny. These paired off with the virgins, mated, and each female then laid a single winter egg. After the harsh winter passed this egg hatched to unlease another population of Amazonian plant lice.

Naturalists soon found that parthenogenesis, or virgin reproduction, is not an uncommon phenomenon in the animal world. Other insects besides the aphids, along with water fleas and rotifers, showed the same propensity for bypassing the male in reproduction. In some species of moths, the bag moth *Solenobia triquetrella* among them, two separate and distinct races exist, one completely parthenogenic and the other strictly bisexual in its reproductive habits. Some of the stick insects, some of the bees, wasps, and their cousins as well as certain centipeds and millipeds were found to have gone even further in their evolution, dispensing entirely with males so that the population and race was entirely female in sex.

The ovists were delighted with this news, for to their mind it proved conclusively that Harvey was right in maintaining that the egg was the essential ingredient in all reproduction and that Leeuwenhoek was wrong. Theologians and ecclesiastics were likewise delighted with this evidence for the possibility of Mary's virginity in giving birth to Christ.

The case of aphid parthenogenesis put the homunculists with their backs to the wall. These observations seemed to reduce the role of the sperm and semen to nothing more than a trigger for development of the egg, if that much remained for it. The more desperate the situation became the more exotic the solutions proposed by Leeuwenhoek, Hartsoeker, and other defenders of male supremacy. Perhaps the atmosphere contained enough floating semen to fertilize the supposed-virgin aphids during the spring and summer. More daring armchair naturalists suggested that any water or moist air could easily contain male seed from countless species or animals so that even higher animals and man himself might on occasion conceive virginally.

Such an armchair observation may seem ridiculous, but considering the state of the natural sciences at the time and of man's understanding of reproduction, it is not unexpected that even serious scientists and philosophers were involved in discussions and reports of mammals and even women giving birth to offspring virginally conceived without male concourse. The cynics, especially the defenders of the ovist theory, had a field day.

In his slim volume, *Can Man Be Modified?*, the French biologist Jean Rostand drew on some of his father's Cyranesque wit in recalling a lampoon of parthenogenesis published in 1750:

Many couples riven by conflict return to harmony, trust, and mutual respect! May slandered women bid gossiping tongues be silent! A woman living apart from her husband can nevertheless present him with the joys of fatherhood, fruit of a boating trip or a country walk. The woman only wanted to get a breath of fresh air, and the organic molecules needed only a draft, a slight breeze, to establish themselves in her. A widowed lady has children, and gossip names their fathers. But there is no need for any such assumption: she merely breathed the air and the molecules crept into her along with it. A girl becomes a mother before marriage: again only the air she inhaled is to blame. Why should humanity be in a worse plight than ordinary insects? . . .

Much has transpired in experimental biology since the lampooning of virgin mothers in 1750. The skepticism remains and rightly so, but both scientists and laymen are less ready to toss the possibility off as a rather naïve joke.

In 1896 Oskar Hertwig and his wife found that a little strychnine or chloroform added to the seawater containing some sea-urchin eggs would set them off developing without the benefit of any contact with sperm. Three years later one of the great founders of experimental embryology, the Frenchman Jacques Loeb, startled the scientific community by confirming and extending the Hertwig venture in fatherless reproduction. Professor Loeb, who later came to the United States to work at the Woods Hole Marine Biological Station on Cape Cod, became an instant *cause célèbre* in both popular and scholarly journals. "When I announced the successful chemical fertilization of sea urchins, the almost unanimous opinion was that I had been the victim of an illusion, and at first I myself was afraid that I had been mistaken."

Loeb's experiment was simple enough and can easily be repeated today by any neophyte to biology with a few elementary instructions and practically no equipment. A female sea urchin is placed upside down over a glass dish and shocked, either with a mild electrical current from a six-volt dry cell or by an injection of a mild potassium chloride salt solution. She will then spill her eggs into the dish, where sperm, collected from males in the same way, can be added to imitate fertilization as it occurs in nature. The result of such artificial fertilization is normal larval sea urchins. In Loeb's experiment the males were left in their aquarium and the eggs merely subjected to the physiological shock of a hypertonic salt solution; a little extra salt, magnesium chloride, was added to the seawater containing the eggs. The result was what appeared to be normal activation of the eggs, cleavage, and healthy plutei or larval sea urchins.

With Gallic sarcasm the staid *Annales des Sciences Naturelles* christened Loeb's parthenogenic progeny "chemical citizens, the sons of Madame Sea-Urchin and Monsieur Chloride de Magnesium." Nevertheless artificially induced parthenogenesis was now a reality in man's growing control over animal reproduction. When Dr. A. D. Peacock summarized the research on parthenogenesis at a 1952 meeting of scientists in Belfast, he reported 371 different ways of producing sea urchins *sans père*. These included 45 different kinds of physical shocks from simple shaking in a glass vial to more esoteric thermal shocks, 93 chemical ways, 64 biological, and 169 combinations of the foregoing, all of which successfully trigger the development of fatherless sea urchins. This listing does not include a considerable number of variations of the basic 371 methods.

Once the sea urchin had been introduced to the joys of paradaisic reproduction it was not long before another Frenchman set out to extend the pleasures of Eden to the experimental embryologists' second favored guinea pig, the frog. Frog eggs proved just as amenable to fatherless generation as the sea urchin but because of slightly different structures in the egg, the amount and distribution of yolk material in particular, Eugene Bataillon had to devise new techniques. Instead of shocking the female to obtain eggs, Bataillon removed several pituitary glands from adult frogs and injected these into the body cavity of a female, some still winter evening while her sisters were still hibernating, some months before the onset of her normal breeding season. The hormones contained in the crude pituitary injection trigger the female's reproductive system so that within forty-eight hours she is ready to release her burden of a couple of thousand eggs. In nature the pressures of the male embrace give a clue to the female so that she releases her eggs just as her mate is releasing his semen into the pond water. In the laboratory it is easier to strip the eggs by hand and expose them to diluted macerated testes from several males. Fertilization can be close to 100 percent if you are skilled in the technique. College and high school students find it fascinating and thrilling to midwife a couple of thousand tadpoles into existence, but even more fun to serve as "father" to a batch of parthenogenic tadpoles. The percentage of success is far lower, usually around 5 to 10 percent, but the accomplishment is on another plane of the esoteric.

To "father" some frogs, you very gently and carefully roll some stripped eggs on blotting paper to remove the three layers of jelly and expose the naked egg. Then you spread a thin layer of blood from a nonovulating female over the naked eggs and carefully prick the surface of each with a very fine glass needle.

For some as yet unknown reason artificial parthenogenesis is possible with the frog only when some foreign protein is introduced into the egg by the needle.

Only a small percentage of parthenogenic eggs develop into healthy tadpoles and even fewer have been raised through metamorphosis to sexual maturity as frogs. Since the egg has only half the chromosome complement of the normal adult cell, the condition of the fatherless tadpole is very intriguing. Parmenter examined many such tadpoles and found some to have just what he expected, half the regular number of chromosomes. But these did not undergo the changes of metamorphosis to become adult frogs. They remained tadpoles and eventually died. Those tadpoles which metamorphosized into frogs somehow restored the normal diploid set of paired chromosomes in all their cells. Two theories have been offered in explanation of this mixed-up condition. You may recall that a frog egg, like a human egg, releases the second polar body when activated by the entrance of the sperm or some agent, chemical or physical. That polar body contains a complete set of single chromosomes, so that if for some reason its expulsion is suppressed and these chromosomes remain in the egg, the egg will then have a double set of chromosomes, normal for a *fertilized* egg. Some scientists feel that this is what happens when frog eggs are triggered to parthenogenetic development. Others suspect that the polar body is expelled as usual, but that after the haploid set of chromosomes in the egg duplicate in preparation for the first cell division, something suppresses the cleavage of the cell proper into two daughter cells. This would leave the original cell with a double set of chromosomes, again normal for the fertilized egg.

After Bataillon's success with substitute fatherhood in 1910, several species of clams, mollusks, and segmented worms were enticed back to Eden by chemicals or the scientist's fine glass needle.

Thus far no scientist has been able to induce birds to join the bandwagon, though at least one bird is known to reproduce occasionally by *natural* parthenogenesis, the domestic turkey. Dr. M. W. Olsen and his colleagues working at the United States Department of Agriculture have found that certain strains of turkeys have a *natural* tendency to produce parthenogenetic offspring. By isolating these, Olsen has developed a special breed of parthenogenetic birds. Once the sex of a hen has been ascertained, it is immediately isolated in a wire-enclosed pen so that any possible contact with a tom is eliminated. After thirty to thirty-six weeks, the special hens are transferred to laying houses. Over a nine-year period many thousands of parthenogenetic eggs have been laid. A high percentage of these developed abnormally but some did hatch and survive to sexual maturity. Between 1956 and 1960 sixty-seven embryos, all males, were hatched. Three of these produced live sperm and one even sired some offspring in normal matings. In some recent experiments, Olsen has achieved as high as 41.7 percent parthenogenetic development in eggs laid by his special breed.

Sooner or later it was to be expected that somewhere a biologist would try to induce virgin birth in a mammal. Oddly enough, despite the evident interest such an experiment holds, it was accomplished not by a frontal attack but rather as an accidental outcome of other experiments with the *in vitro* culture of mammalian eggs. Nearly forty years ago the French embryologist Champy discovered among a batch of unfertilized rabbit eggs maintained at body temperature in a culture dish some which had started cleaving even though they had not been fertilized. Extensive research was then undertaken by Gregory Pincus and his associates in the United States. Most of this work was done in the 1930's with rabbit eggs, although some experiments touching on the question were carried out with *in vitro* culture of human eggs. Pincus found that if the eggs were simply left for a day or two in ordinary tissue culture medium—blood plasma plus embryo extract—some of them would be activated. Some eggs even went beyond extrusion of the polar body and started cleaving. The chemical treatments used with sea urchins were tried but without success. However, a good temperature shock, particularly cold, proved an effective trigger.

Once activated the eggs could not continue their development for very long in the tissue culture apparatus and media then available, so that in some experiments Pincus transferred the activated eggs to the fallopian tubes of pseudopregnant rabbit does. There development could continue with quite a few embryos reaching the blastocyst stage. In one experiment by Chang 18 percent of the rabbit eggs that were cooled for twenty-four hours at 10°C. were activated. Among the many embryos which Pincus started off fatherless, only two completed a full term in their substitute mother's womb. One of these was stillborn, but the other came out healthy, alive and kicking. This was in 1939. A year later Pincus and Shapiro operated on a doe rabbit and applied a cold compress to her fallopian tubes which at the time contained unfertilized eggs. One normal live rabbit was born as a result of this thermal shock treatment, achieving the fame of a cover picture on *Life* magazine.

Actually this work with parthenogenesis was peripheral to the interests of Pincus and his co-workers, who were concerned with problems of contraception and the hormonal control of egg fertilization or implantation. Despite the challenging new vista that was opened up by this work, Pincus and his associates stuck to their main concern, the contraceptives, and the work remains unconfirmed and unrepeated to this day.

If this is the situation among the other animals, as far as natural and artificial parthenogenesis goes, is it possible that women may occasionally be victims of a virginal conception? The vast majority of scientists and medical practitioners would agree that it is practically out of the question. Yet few will eliminate the possibility completely.

There are some interesting reports we can cite, even though none of these really answers our question.

In 1944, for instance, a twenty-year-old woman, Emmie Marie Jones, collapsed of extreme exhaustion during a heavy bombardment of Hanover, Germany. Although she denied any physical relations with men, she became pregnant and gave birth to a daughter, Monica. In 1955 some English scientists investigated her case and found no evidence of any differences in hereditary traits between mother and daughter. Genetically, the two appeared to be identical twins. However, a skin transplant, which is ordinarily successful between identical twins, failed to take between the two women. The result: a possibility of virgin birth, perhaps even a probability depending on how substantial you find the evidence, but not any degree of scientific certainty.

We should in passing comment on the sexuality of virginally conceived offspring. Sex in the birds, like the domestic turkey, is

determined *by the female*. Male birds, unlike humans, have two
X chromosomes. Female birds have only one X chromosome, but
no Y or male-determining chromosome. Thus the father's only
contribution toward sex in his offspring is an X chromosome. If
the X-bearing sperm fertilized an X-bearing egg, the offspring is
male; if the egg contains no sex chromosome, the offspring is
female. Thus parthenogenetic birds are all male, unless something
really goes awry during the first cleavage. The opposite is true
of mammals where the chromosome content of the sperm deter-
mines the offspring's sex. Parthenogenetic mammals and humans
then would always be female, since no Y chromosome is involved
in their heredity.

Yet even this generalization might not hold true if you consider
the strange case reported by Dr. Walter Timme at a meeting of
the American Neurological Association. The case involved a
sixteen-year-old Arkansas girl who was being operated on for an
ovarian tumor. As is the custom in such surgery, the tissue removed is carefully examined by a pathologist. In this instance,
signs of live eggs *and* live sperm were found in different regions
of the tumor. With the egg and sperm situated right next to each
other in the same organ, Dr. Timme claimed "there was a great
possibility that they could combine and make a human being."
This would not be strictly speaking a virginal conception since
the offspring would be the product of both male and female germ
cells. The unique feature though would be that the *same* person
contributed *both* germ cells. As Timme stated later, "it was a true
case of where the physical setup was possible for a virgin
birth." The medical profession and experimental biologists have
always been very skeptical about the existence of functional
hermaphrodites among the higher animals and man, though the
earthworm, the sea hare, and other lower animals do combine
both sexes in the same individual.

One is tempted to wonder what meaning the word "virgin"
takes on in a case like that reported by Dr. Timme. Would this
girl still have been a virgin had she conceived without sexual
intercourse?

Several researchers have reported the spontaneous activation
of human eggs maintained *in vitro* where any possible influence
of sperm was definitively ruled out. This often happens as un-

fertilized eggs begin to degenerate. But in the late 1930's Dr. Stanley P. Riemann of Philadelphia's Lankenau Hospital reported his success in mechanically activating the human egg with the prick of a glass needle. Though the eggs released the polar body as a normally fertilized egg would, they lived only eight hours and did not cleave, the usually accepted criterion for fertilization. Some work of Dr. Landrum B. Shettles is even more to the point, for in examining human eggs just after they were removed from their ovarian follicles he found that three out of four hundred of these eggs had "undergone cleavage *in vivo* within the intact follicle, without any possible contact with spermatozoa." If Dr. Shettles' observations are any indication of the natural state of affairs in the human ovary, it would seem that about three-quarters of one percent of maturing human eggs begin a parthenogenetic development even before they leave the shelter of the ovarian follicle for their trip down the fallopian tube to the womb. It would follow that virgin births are a rather common occurrence, in about the same frequency as fraternal twins and twice as often as identical twins occur among white Americans. (American Negroes have a higher rate of both types of twins than Caucasians in North America.) This extrapolation to the frequency of virgin births is a mere educated guess, without any substantial evidence. Dr. Sherwood Taylor, a noted British scientist, has suggested a much lower frequency for human parthenogenesis, estimating one case in ten thousand births.

Granted the human situation today there is absolutely no way to test the validity of these estimates. The majority of women who have babies are married and would hardly be inclined to suspect a virginal conception. With unmarried women the prevalence of premarital sexual intercourse would lead to a similar skepticism of any claim to virginal conception, if not by the woman herself, at least by the doctor she might consult and others.

Considerable interest and speculation was stirred up some years back in Great Britain by a woman scientist, Dr. Helen Spurway, a well-known lecturer and an expert in eugenics at London's University College. In 1955, Dr. Spurway gave a talk on "virgin births," suggesting that in terms of laboratory experiments one in every 1.6 million pregnancies would involve a spontaneous and natural doubling of the egg chromosomes resulting in a parthenogenetic birth.

The reactions were not as skeptical as one might have expected though there was considerable reserve and caution on the part of journals like *Lancet,* a British medical publication. "The possibility that a woman might become pregnant without at least one spermatozoon having entered the uterus is not one which the 'reasonable man' would lightly entertain." Furthermore, the editors of *Lancet* noted, "scientific opinion for several centuries has sided with the reasonable man." If it does occur, then it is extremely rare. The *Lancet* editors suggested the very cautious estimate of a frequency about one-half that expected for the birth of sextuplets, one virgin birth in approximately three billion pregnancies. (The world population today is roughly 3.5 billion.) Several criteria were suggested by both Dr. Spurway and the editors as possible tests for verifying cases of supposed virgin births: successful long-term skin grafts which have been considered strong corroborative proof of parthenogenesis in turkeys, as well as blood and serum comparisons.

As was to be expected the newspapers picked up the theme as a handy tidbit of scientific information ripe for exploitation by the headline media. Even the urbane *Manchester Guardian* jumped at the bait, speaking of "life *without* Father." The *Sunday Pictorial,* a tabloid often enchanted by the potentially sensational, decided to chalk one up on its competition by sponsoring a scientific test of the theory. If Dr. Spurway was correct, then a number of Englishwomen probably had experienced parthenogenetic conception. These could be subjected to rigorous testing. The editors approached Dr. Stanley Balfour-Lynn of Queen Charlotte's Hospital and asked him to direct the testing, supported by a team of distinguished consultants in various fields of the medical and biological sciences. Nineteen women came forth in response to the invitation of the *Sunday Pictorial* articles. Several of them withdrew immediately when it became evident that they did not understand the exact meaning of parthenogenetic conception. In the end, after many tests, one mother and daughter remained who appeared to be identical in their genetic constitution.

Many, including such eminent scientists as Dr. J. B. S. Haldane, the husband of Dr. Spurway, however remained skeptical despite the indirect evidence. In cases such as these, where one is trying to prove that something did not happen, the only hope is to prove that the contrary did in fact occur. In simple terms, we can never prove that a child was conceived without any sexual contact or paternal contribution when we examine that child after birth. If, however, we find that the child's genes as expressed in its blood, skin, and serum types are not identical with those of its mother, then we can safely assume that some hereditary contribution besides the mother's was involved in its conception. Thus we can *eliminate* certain candidates for virgin birth, though we cannot hope to prove in any individual case that a child was conceived without a genetic father. Rigorous proof, as Dr. Balfour-Lynn has pointed out, is impossible though he found the evidence favoring a virginal conception in this one instance entirely consistent with what would be expected from serology and other tests of both mother and her parthenogenetic offspring.

It is very unlikely that parthenogenetic reproduction will ever become anything more than an experimental tool for the geneticist and the developmental biologist. Students of embryology will continue to delight in "fathering" frogs and sea urchins, but when you come down to the basic realities of life, even if we can sometime in the future trigger virginal conception in humans, will it ever become anywhere near as popular as artificial insemination, surrogate mothers or artificial wombs? Most unlikely, I suspect. In fact, I would expect very few if any applications for human virginal conception, whether natural or artificially induced. At least in terms of human reproduction, its morality and its future trends, this aspect of the new embryology holds little more than speculative interest. Our discussion of it, however, will prepare the way for other more relevant modifications in asexual reproduction.

The embryologist has already found some intriguing variations of the theme, in some cases ways to bypass the egg and sperm completely. One innovation, provided by the now classic experiments of Drs. Robert W. Briggs and Thomas R. King, has absolutely no parallel in nature though it will evoke some engaging and serious controversies in the area of human morality and customs.

These experiments culminated in research by Briggs and King at the Institute for Cancer Research in Philadelphia in 1951 and 1952. But the origins of the idea can be traced back to 1938, when one of the monumental figures in embryology, Hans Spemann, suggested an experiment which at the time seemed to him "somewhat fantastical." The proposal was to remove the nucleus from an egg and replace it with the nucleus of some other cell, preferably an adult cell. "This experiment might possibly show," Spemann hoped, "that even nuclei of differentiated cells can initiate normal development in the egg protoplasm." (Which brings us back full circle to Messenger's remarks at the beginning of this chapter.) Spemann's creative genius was already geared in this direction, for he had already managed to constrict a fertilized salamander egg with a fine loop of baby hair so that the nucleus was pushed into one side and unable to pass across the tiny bridge of cytoplasm which connected the two halves. Eventually, after three or four divisions, the nuclei had become small enough so that one could pass through the bridge and enter the "virginal" undifferentiated cytoplasm on the other side. In the process of dividing several times, it was thought that certain genes would be turned on and others turned off in the individual daughter nuclei which would then develop along certain specialized lines into adult organs and structures. If a nucleus had developed along a certain channel or line, would it be able to reverse its specialization, return to the embryonic condition, and trigger the normal development of an enucleated egg into a whole animal? The nuclei available in Spemann's delayed nuclear supply experiments had not differentiated very much, not nearly as much as the nucleus in an adult liver or kidney cell, for instance. Hence, Spemann's "fantastical" plan for transplanting adult nuclei into enucleated eggs. Little could be learned about the process of cell differentiation which underlies the whole question of development by transplanting an adult liver nucleus into an adult kidney cell because the new host cytoplasm would be too alien to the transplanted nucleus to engage in physiological exchanges and reactions. The relatively undifferentiated egg material should, however, make an ideal host for a foreign nucleus.

The two scientists began by testing the consequences of removal of the egg nucleus on the capacities for development in the cytoplasm of the egg. Clear evidence emerged that the cytoplasm is helpless without its nucleus and that only the transplanting of an "undifferentiated" nucleus from a closely related zygote will allow it to develop. What happens, however, when the nucleus of a differentiated cell is used for the transplant? Will further development be normal?

To answer this basic question Briggs and King developed two highly sophisticated and skilled techniques: a 99 percent effective method for enucleating the frog egg and a way to remove and transplant the nucleus of a differentiated cell of an older embryo without damaging it.

First, they pricked a frog egg with a *clean* glass needle so that its outer coat, the cortex, rotates within ten to fifteen minutes and the egg nucleus appears as a small black dot in an opaque depression on top of the egg. If the nucleus is left alone, abortive cleavage furrows appear in five to ten hours. But the nucleus can be removed by inserting a glass needle beneath it and gently flipping it out of the egg. In this case, no cleavage furrows appear, at least for fifteen hours.

With the triple jelly coat removed from the enucleated egg chemically or by gentle rolling on absorbent paper, the hosts for this transplant are set in small depressions in wax-bottomed watch glasses with a special culture medium devised by Drs. Niu and Twitty. In the center of the dish, a one- or two-day-old donor embryo rests, collapsed like a half-deflated balloon. Beneath the dissecting microscope, two glass needles are manipulated freehand like chopsticks to probe beneath the surface of the embryo and remove one of its subsurface cells. An undamaged surface on the embryo serves as an operating table on which the cell can rest as the technician prepares to remove its nucleus. The cell is first drawn up into a clean micropipette with an inner diameter slightly smaller than that of the cell. If the cell membrane is torn open in the process, the cytoplasm will escape and the nucleus will be damaged or killed. On the other hand, if a slight break is not made in the cell membrane, the nucleus will remain isolated from the host egg and the transplant will fail. Just enough of a tear—not too much, not too little! The surface of the cell must be broken but the contents left undisturbed. Finally the micropipette with the cell at its tip is gently pushed deep into the enucleated egg and deposited there.

It is remarkable that, even in their earliest experiments, Briggs and King got more than half (57 percent) of their recipient eggs to cleave and the majority (34 percent) of these to develop into complete embryos. In these early experiments the donor nucleus was taken from blastula embryos roughly fourteen to twenty hours after fertilization. Nuclei from a later stage, the early gastrula, when the three basic tissues are forming some twenty-six hours after fertilization, were less successful, 38 percent cleaving and about 15 percent forming complete embryos. Several theories were formulated to explain the difference and months of hard work were put into testing these. When it was suspected that the older nuclei might be more easily damaged during the operation, Briggs and King found that a calcium-free solution containing the digestive enzyme trypsin would separate the gastrula cells with much less damage. With this technique the percentage of cleavage and full development achieved was improved several fold over that achieved with mechanically isolated nuclei.

Again many months of hard detail work passed before an overall picture began to emerge. One important fact which came to light was a difference in the capacity of nuclei from different regions of the same animal to produce normal embryos when transplanted. Of nuclei transplanted from cells, which in a twenty-six-hour-old embryo will normally form the superficial ectoderm or skin, 41 percent cleaved and 80 percent of these developed into fully normal tadpoles. If a cell from the future intestinal region is used, again 40 percent will cleave but only 20 percent of these will hatch four or five days later. Half of these intestinal nuclei transplants end up aborting with abnormally small central nervous systems, reduced sense organs, and other deficiencies of tissue derived from the superficial ectoderm. This told Briggs and King a very important fact: nuclei in the intestinal cells apparently undergo differentiation and specialization very early so that while they contain all the original genes, some of these have been turned on and others turned off very much like switches on a master control panel. Once a certain sequence of genetic decisions along a certain developmental pathway has been made inside a nucleus, it becomes increasingly more difficult to reverse the controls and start off in another direction, or, as Briggs and King were trying to do, start the nucleus off developing into a complete whole embryo with all its parts and organs. Nevertheless, some nuclei were able to reverse gears. The genes which had been switched off in the intestinal cell because their specific protein and enzyme products were no longer needed by the digestive cell were somehow switched back on by the undifferentiated cytoplasm of the host egg. Even so, the more a cell specializes in a certain direction the more difficult the reversal becomes.

Viewed with the critical eye of the scientist these early results, informative as they appeared, were wide open to the objection that the intestinal nuclei were damaged and not necessarily genetically specialized. An answer to this doubt could be uncovered if it were possible to clone individual animals after the original transplantations. "Clone" comes from the Greek for throng and is used by scientists today as a shorthand label for a population of individual organisms, plant or animal, which originate from a single parent, asexually. Each individual in the throng or clone, because he receives his genes from only one parent, is an identical twin of his parent and of all his brothers in the clone.

With their nuclear transplant technique now perfected, Briggs and King set to work producing clones of several generations. After a dozen nuclei from the intestinal region of a single gastrula tadpole had been transplanted into a dozen enucleated host eggs, one of the new generation was selected at random as donor of intestinal cells at the same age as the original transplant for a dozen more transplants. This was then repeated with a volunteer from the third generation, the next, and the next. When specific nuclei were transplanted in this way over several generations with thousands of offspring it was found that they indeed had specialized genetically, for the cloned offspring were very similar to each other. All the members of a clone either aborted their development at about the same stage and with very similar abnormalities, or generation after generation in the clone matured normally.

BRIGGS AND KING	J. B. GURDON
(Experiment with Grass Frog) *Rana pipiens*	(Experiment with African Clawed Frog) *Xenopus laevis*

1. Nuclei taken from endodermal (intestinal) cells of a donor in the late gastrula stage of embryonic development.

1. Nuclei taken from the intestinal cells of a tadpole which normally has one nucleolus in the nucleus of each cell.

2. Single nuclei are injected into eggs which have had their own nuclei removed mechanically.

2. Single nucleus is injected into an unfertilized egg from a two-nucleolated strain which has had its original nucleus destroyed by ultraviolet radiation.

3. Individual animals in this first clone will halt their development at different stages, because, though genetically identical, their various gene complexes are now functioning differently due to the donor cells' slightly different locations within the presumptive gut of the donor blastula embryo.

3. Allowed to develop to the blastula stage, genetically identical nuclei are isolated and then transplanted into ennucleated eggs, giving rise to a clone of asexually produced identical twins.

4. One of the original recipients of late gastrula nuclei is disassociated and its intestinal cells provide nuclei for a first blastula nuclear clone.

4a. Most of this first clone are allowed to develop as far as they can, in some cases to full adult frogs. Each member of the clone will halt its development at approximately the same stage, for they are identical twins.

4b. One cloned twin is sacrificed as a control at blastula and several of its nuclei transferred to a second batch of ennucleated eggs which will match in development their "first cousins once removed."

5a. All the embryos in this first true clone will develop to the same stage, sometimes as adults.

5b. One of the first blastula clones is sacrificed and its endodermal nuclei transferred to ennucleated eggs for a second blastula clone. This cloning can be continued indefinitely with the individuals in all the blastula clones being identical twins to each other.

5. This cloning can be extended indefinitely over many "generations" with the "progeny" of the fifth or sixth transfer being identical twins of those in the first generation.

NOTE: The presence of nuclei with only one nucleolus in the transplant progeny proves the original nucleus has been removed and actually replaced by the donor nucleus.

FIGURE ONE: CLONING EXPERIMENTS

A decade after Briggs and King began their work in earnest on nuclear transplants with the grass or leopard frog, Dr. J. B. Gurdon pushed cloning one step further. Gurdon's guinea pig at Oxford University's Department of Zoology was *Xenopus laevis*, the South African clawed frog, which was once used in human pregnancy tests. Though Gurdon used the basic Briggs and King technique he introduced two vital modifications. Unlike the *Rana* egg used by Briggs and King, the *Xenopus* egg does not take well to mechanical enucleation, so the egg chromosomes were destroyed by a tiny beam of ultraviolet light. The other change was in the donor. Instead of using nuclei from larval cells, Gurdon tried the intestinal cells of adult frogs where specialization was complete. Despite the use of these much older, more differentiated nuclei, the results were very much parallel to those of Briggs and King, proving that it is possible to transplant the complete genetic code of one individual into literally dozens of hosts, thereby producing asexually a whole population of identical twins extending over several generations.

Not only does this experiment demonstrate that reproduction can now bypass the sexual germ cells, egg and sperm almost entirely, it also contains the seeds for potential immortality for the individuals cloned. Because the same genetic code is serially transplanted over several generations, the cloned individual is in fact perpetuating his genetic self without the slightest change into offspring that are really not offspring. The individual in the second generation has, it is true, only one parent, a "FATHER-mother" or a "MOTHER-father," depending on the sex of the donor. But in sex and in every other inherited trait he is the identical twin of his "parent." Genetically he *is not* an offspring; he *is* his own parent reincarnated in some new cytoplasm.

The frog egg is about an eighth of an inch in diameter and fairly easy to operate on under the low-power magnification of a dissection microscope. But the mammalian egg is only one-twelfth that size, about a hundredth of an inch in diameter, the size of the period at the end of this sentence. Today a skilled technician can transplant an adult nucleus from some organ or from the skin into a frog egg. Is it possible to do the same with a mammalian egg? Can we enucleate a human egg and transplant into it one of my skin's cells to produce my identical twin minus thirty-odd years? So far no one has succeeded in this, or even attempted it. Dr. Gurdon himself plays down the possible application of his technique to man as idle speculation and sheer crystal gazing. Yet other scientists are not as pessimistic about the application. Already our technology of micromanipulators is speeding to the point where the precision required for the enucleation and nuclear transplant of a mammalian egg will be a reality. Specialists like Dr. M. J. Kopac at New York University have already gone far along the road to such microsurgery. Dr. Kopac has, for instance, successfully transplanted the nucleolus from one cell to another.

So far in this chapter we have explored some indications that man has already the potential for natural virgin birth, that imitating Pincus' thermal shock of a rabbit egg we may sometime in the coming generations be able to trigger artificially and at will human parthenogenesis, and even more interestingly transplant mature adult nuclei into enucleated eggs. All these techniques highly modify the normal processes of fertilization and conception as we know it in our sexual world, but they still remain dependent in part on the egg, even if its nucleus is removed. We can now turn our attention to the possibility of cloning or reproducing asexually organisms without any reference at all to the sexual cells. Forget completely the egg and sperm and let us discuss the plain possibility of reproducing plants, animals, and *perhaps* even human beings in a truly asexual manner.

The idea we propose is not a new one in scientific circles. In 1902 a middle-aged Austrian plant physiologist by the name of Gottlieb Haberlandt tried his hand without success at culturing plant cells *in vitro*. Commenting on the situation at that time and the distinct advantages he saw in learning how to culture both plant and animal tissues in controlled environments, Haberlandt admitted:

> There has been, so far as I know, up to the present [1902], no planned attempt to cultivate the vegetative cells of higher plants in suitable nutrients. Yet the results of such attempts should cast many interesting sidelights on the peculiarities and capacities which the cell, as an elementary organism, possesses. They should make possible conclusions as to the interrelations and reciprocal influences to which the cell is subject within the multicellular organism.

Haberlandt was not apparently aware then of the possible implication of tissue culture for asexual reproduction of whole animals or plants. But it was there, even in 1902, in the minds of some.

The basic problem to any culturing of cells *in vitro* is the proper medium. Wilhelm Roux had kept the neural plate of a chick embryo alive for a few days in the 1880's by using a warm saline solution, but the first major steps in tissue culture had to await the work of three American scientists some years later.

The American troika consisted of Ross Granville Harrison, Montrose Burrows, and Alexis Carrel, with an assist from Charles Lindbergh.

Born in Germantown, Pennsylvania, in 1870, Harrison graduated from Johns Hopkins University at the age of nineteen after majoring in biology and mathematics. With a doctorate from Hopkins in 1894, he went on to Bonn and earned an M.D. degree in 1899. Returning to the States, he accepted a teaching position at Yale University, and in 1907 published a paper which today is generally considered the birth of the science of tissue culture. That paper reported Harrison's success in growing neurones, nerve cells, outside the animal body. He had taken different tissues from frog tadpoles and placed them in drops of clotted frog lymph. Cells which grew from the edge of the neural plate proved to be normal nerve cells.

Alexis Carrel, one of the most colorful figures in modern medicine, heard of the remarkable success of Harrison and sent Montrose Burrows to work with him. In three years Burrows picked up a very valuable background and much information on techniques. In the process he also made his own contribution by showing that chicken plasma was far better as a culture medium than frog lymph. Burrows returned to Carrel's laboratory in 1910 at the Rockefeller Institute in New York City. It was Burrows who actually coined the term "tissue culture."

Harrison's accomplishment in tissue culture was so extraordinary that the 1917 Nobel Prize Committee recommended him for that honor, but in the end no prize in medicine was given that year. In 1933 his name was again recommended, but "in view of the rather limited value of the method and the age of the discovery" the suggestion was turned down. So limited was the value of Harrison's technique that it took twenty-one years before Enders, Weller, and Robbins used the method to culture the polio virus, preparing the way for the discovery of the Salk vaccine. For this application Enders, Weller, and Robbins received due recognition in the 1954 Nobel Prize.

Trained in France as an experimental surgeon, and later the recipient of the 1912 Nobel Prize, Dr. Alexis Carrel had a long-standing interest in wound healing and tissue proliferation. He felt very strongly that the only way to understand how tissues heal when injured would be to investigate the growth of cells under experimental and controlled conditions which eliminate the complications and variable factors of normal *in vivo* conditions. When Burrows returned from Hopkins in 1910, the two set their efforts on a breakthrough in tissue culture. They began by keeping Rous fowl sarcoma cells, a type of cancer, alive for two generations *in vitro*. Then came some work with various tissues from cats and dogs. Their success was greeted by almost universal skepticism, particularly in Europe. Provoked by this response, the brilliant Carrel set out to prove beyond any doubt their success with tissue culture. In 1912 he placed some embryonic chick heart tissue into *in vitro* culture. Fifteen hundred generations later, in 1921, the cells were still growing and multiplying in the test tube. They were eventually passed on to the Lederle Laboratory and maintained in excellent condition until 1946, two years after Carrel died.

Carrel's many other interests included basic problems of organ transplantation, wound infection and blood clotting, the invention of new surgical instruments, and the discovery of a new suture technique. In 1930, three years after his transatlantic solo flight, Charles Lindbergh joined Carrel. Lindbergh knew nothing about medicine but his inventive mind explored many fields. A relative with "heart lesions" was interested in a possible surgical cure. This prompted Lindbergh to conceive and design a perfusion apparatus which this surgery would require. Together Carrel and Lindbergh perfected this new piece of equipment and soon kept a cat thyroid alive for eighteen days. This took place in 1935, long before scientists began to work on artificial wombs.

The development of effective tissue culture methods was the keystone without which scientists could not hope to achieve anything in the way of Haberlandt's dream of asexual reproduction by cloning individual cells. The man who has carried this technique furthest and fastest along the road to reality is Dr. Frederick C. Steward, a chemist by training with a botanical bent. Director of Aircraft Equipment for the British Ministry of Aircraft Production during World War II, Professor Steward is now director of the Laboratory for Cell Physiology at Cornell University in upstate New York.

With initial experiments performed on the mature tissues of the carrot root, his hope was to "reactivate" these resting cells so that they would grow, possibly into mature plants. The technique is extremely taxing in its precision and care for detail. All the operations from the first removal of cells from the carrot root till the new plant is well on its way must be done aseptically and with all the precautions of a surgical operation. After removal the cells are placed in sterile glass tubes, stoppered with cotton, and attached to a rotating arm which rocks the tubes back and forth so that the cells are alternately submerged in the culture medium and exposed to air. The medium is a basic artificial solution containing all the raw materials needed for plant growth plus coconut milk. The embryonic plant in a seed, be it corn, pea, or bean, is surrounded by endosperm, a specially adapted food material which carries the embryo plant through its earliest development until it develops roots to absorb food from the ground. Coconut milk is a liquid endosperm material and very handy for experiments in the cloning of plants. In a mere twenty days, Steward's resting carrot cells became "alive," increasing some eighty fold in weight. Until he decided to try the coconut milk, results had been very unsatisfactory. The addition worked wonders. "It was as if the coconut milk had acted like a clutch, putting the cell's idling engine of growth into gear. . . ."

The tempo of the experiments was accelerated by this early success. Sweet potatoes, parsnip, Jerusalem artichoke, and white turnip were tried. White potatoes, onion and maple cells were among the more disappointing subjects used in attempts to trigger simple growth.

Soon the experiments were focusing strictly on what could be labeled cloning of plants. Isolated cells were cultured alone.

Freed from the restricting influence of the organized root tissue, the individual carrot cells now began to grow in a variety of ways. Some grew to giant size, with many nuclei, and showed active protoplasmic streaming. Others put forth tubular outgrowths that became filaments and then divided. Some formed small, uniform buds, almost like the growth of yeast cells. Eventually some of them partitioned themselves into a clump of cells resembling an early carrot embryo. . . . Transferred to an agar medium containing coconut milk, such clusters of cells would then grow shoots opposite the root; that is, they became plantlets very similar to the normal embryo of a carrot plant. Such plantlets, carefully nursed along on the agar medium and then successively transplanted to vermiculite and soil, would eventually grow into mature carrot plants, with normal roots, stalks, flowers and seeds. In short, Haberlandt's old dream was finally realized: complete, normal plants were grown in culture from free (asexual somatic) cells. Cloning!

Several herbaceous plants have since been cloned, tobacco, endive, and parsley. But until 1969 the higher plants, particularly trees, resisted this type of treatment. Then, after six years of chemical and environmental manipulations, a group of scientists at the Institute of Paper Chemistry in Appleton, Wisconsin, announced the first cloning of a tree. The Institute, a graduate school and research center specializing in research and technology related to the pulp and paper industry, is affiliated with Lawrence University nearby. Its facilities offered a unique opportunity for Dr. Martin Mathes, now at William and Mary College, Dr. Lawson Winton of the Institute's Genetics and Physiology Group, and Dr. Karl Wolter of the Forest Products Laboratory in Madison. Their subject was a triploid aspen, an unusual tree with three sets of chromosomes instead of the customary two, a characteristic which gives it some very superior properties in terms of pulp and paper production. Starting with plugs one-quarter of an inch long and one-eighth of an inch in diameter taken from the mutant tree's cambium, a shiny, sticky growth layer just beneath the bark, Dr. Winton's team manipulated the light, heat, moisture, vitamins, nutrients, and hormones into just the right combination to trigger growth of the first artificially produced tree.

Plants and trees are one thing, animals and humans are another, and most scientists are very cautious in predicting when they will breach the barrier between the two worlds and start cloning animals. The scientists may be cautious, but few if any will say that cloning of animals and humans is beyond our potential technology. The only question appears to be *when* this development will become a reality. When it does, the problems we will face are frightening, so perhaps we should start asking some serious questions now and learn to cope with them.

For the lexicographers cloning is likely to produce some mind-rattling puzzlers. How will the new dictionaries define "twin" when this refers to two individuals separated by fifty or a hundred years, born in two entirely different cultures? How does one define "father" and "mother" when cloning is the mode of reproduction? Is there such a thing as an "offspring" when you are dealing with serial cloning? What does "individual" mean in this context? Is there such a thing?

For those concerned with ethics and human morality cloning is likely to produce even more serious questions. Stanford's Joshua Lederberg summed it up beautifully when he said that cloning places mankind "on the brink of a major evolutionary perturbation." Perturbation, the dictionary tells us, is used by astronomers to indicate a major irregularity in the motion or orbit of a heavenly body caused by some outside force. This is exactly the impact cloning is liable to have on the smooth course of human sexuality which for untold eons has orbited around the twin suns of the male and female.

Cloning, Jean Rostand tells us, "would in theory enable us to create as many identical individuals as might be desired. A living creature would be printed in hundreds, in thousands of copies, all of them real twins. This would in short be human propagation by cuttings, capable of assuring the indefinite reproduction of the same individual—of a great man, for example!" The popular media has blown this whole subject out of its proper perspective with enticing headlines like "Want your own Britt Ekland? Make yourself one," beside a cloned septuplet of that young lady appropriately attired in a red jump suit with a navel-plunging neckline. The gross sensationalism and oversimplification inherent in stories headlined by such captions as "Einsteins from cuttings" and "J'aime Mozart XXIII" are hardly justified by the plea that we must educate the general unsophisticated public to the advances and threats of science. Such journalistic headlines and stories abound, but they are not primed for an educational purpose which is, if at all, only an accidental outcome of the scanty substance and information revealed by these articles.

If Gordon Rattray Taylor, the respected British science writer, is correct in estimating the first cloning of animals for the last decade or two of this century and the first cloning of humans shortly thereafter, it will not be by such simple methods as a gardener taking a cutting from his favorite rose or Rostand's "people farming." When, and if, cloning is extended into the animal kingdom, it will evolve out of years of patient, frustrating research by thousands of scientists around the world. It will entail tremendous attention to technologies of physiology, embryology, and biomedical engineering. The accomplishment will be as Lederberg pointed out a biological and medical tour de force.

Granted this caution, we might legitimately ask whether even the most ardent devotee of the muses could take twenty-three Mozarts, a dozen or more variations of Beethoven's Choral Symphony, or a polychromed gross of Berlioz' Symphonie Fantastique. If the scientific community of lesser mortals still finds one Einstein more than they can handle, what would it do with a dozen such geniuses all cut from the same genetic mold?

In terms of our present technology and the foreseeable future it will be much easier to clone a population from embryonic or near embryonic cells than to use cells from the skin of a proven genius in his fifties or sixties. Despite Gurdon's success with the highly differentiated intestinal cells of adult frogs, it is more likely that the first human clones will be derived from less specialized and more embryonic cells. This creates a problem in our selection program. One of the prime purposes behind the suggestion that we clone human beings is the desire to propagate rare and desirable genetic combinations within the human species. If we must resort to embryonic or very young donors, then we can only guess who might provide the best genetic constellation. In the end we might find ourselves asking everyone to donate a piece of skin during their childhood for cold storage and offering very slight odds for the individual that some forty or fifty years later his cells might be cloned.

J. B. S. Haldane, whose skepticism of virgin birth we noted above, took an almost opposite tack in dealing with the future of cloning. On this related subject Haldane was close to enthusiastic. Somewhat eccentric though one of the most brilliant and practical of modern scientists, Haldane accepted cloning as a definite possibility in man's growing technology of reproduction. He suggested that most clones should be made of people in their fifties at least, from those who have proven themselves outstanding in different areas of science, music, art, literature, scholarship, and other intellectual and cultural endeavors. With such achievements the test of time is the hallmark of quality. Recognition of artistic genius often comes to the truly great late in life, sometimes even after death, as for instance many of the painters of the last century. In the field of physical endeavors, however, where the criteria of success are readily evident Haldane suggested cloning in the prime of life. Few would dispute the desirability of a forward wing composed of three Glen Halls or Gordie Howes, or an outfield of three Babe Ruths, or a paddock of Man o' Wars. A similar process could be applied to the field of entertainment with equally incredible results. Picture if you will, as *Esquire* magazine did recently in its sensational style, a chorus of Mahalia Jacksons, a jazz sextet of Ella Fitzgeralds, or a vaudeville production starring a Charlie Chaplin octet. It is not inconceivable or beyond the prospects of the near future to imagine the Fred Astaire of the year 2000 deciding to clone a few selves and train them properly in the art of dance before the original passes on.

The possibilities are even more interesting if we should be able to combine both cloning and embryonic fusion as mentioned in chapter three. Individual cells from Barbra Streisand could be fused with cells from a leading Music Hall Rockette and a real superstar would be born or decanted. Similar creative combinations might be advantageous in athletics and other fields.

On the other hand, the golden prospect of immortal genius raises the specter of prejudice and inequality through selection. Will everyone be allowed to clone themselves if they so desire? To allow this would be to undermine the whole project of improving the human genetic pool. Some kind of control or selection will be a practical and unavoidable necessity. Lord Rothschild, in 1967, foresaw the distinct *probability* of some sort of Commission for Genetical Control for the human race. This commission would examine applications for cloning from everyone and then license only a few while vetoing the majority. This, as Rothschild admitted, would bring into daylight the egomaniacs and egotists everywhere who when thwarted would likely resort to illegal and devious maneuvers of every sort to circumvent their being blacklisted. Faced as we are with a population crisis, cloning might add to the explosive numbers unless controlled by some effective legal agency. Besides limiting the right to clone oneself, a government might also find it necessary to outlaw all sexual reproduction to the point of introducing hormonal contraceptives into the water supply or requiring an annual contraceptive vaccination of all men or women over the age of ten.

"We are likely to face quite pressing social and personal problems," when cloning becomes a reality, Gordon Rattray Taylor warns, "and how far we go towards the eventual problems will depend on how far we can integrate such methods into our culture." Some governments may decide to outlaw completely any use of cloning for humans, much as they now try to outlaw abortion or contraceptives. Others may wholeheartedly endorse and even encourage it as *the* most desirable mode of reproduction, much as some countries now have birth control or abortion programs as a national policy. Instead of a nuclear arms race, future generations may find themselves involved in a cloning race to produce ever more brilliant teams of scientists, more courageous soldiers, or more docile public servants. But with the East more likely to plunge ahead with this human engineering project, as Taylor points out, "the problem which would face the western civilizations would then be whether to compete or perhaps face extinction—culturally if not militarily."

In his fascinating though now dated broad survey of the whole biological revolution, Taylor highlights some of the questions we will have to face in the near future:

From whom should one take the cell from which a hundred thousand duplicate progeny will be bred? (With what intensive scrutiny society will look at the first products of such an experiment!)

How will the members of this new caste themselves feel about it? Will they be an elite group, only permitted to marry among one another in a sort of mass incest? If not, if they marry freely with uncloned people, the virtues of their carefully selected heredity will be dissipated—though of course there will be some upgrading of the general level by dispersing the desired genes. Or will they be forbidden to marry outside the clone, perhaps?

Among our criteria for cloning should we include some allowances for special vocations and their requirements? Should we perhaps consider the cloning of special mutant human forms as Haldane has suggested? Under ordinary earthbound conditions phocomelia, a mutation which leads to nondevelopment of limbs, is a terrifying prospect for any child. But in a space capsule the lack of legs would be a distinct advantage, especially on long voyages. Is it conceivable that future generations of space travelers may clone their astronauts from a special mutant combining a lack of legs with a genius for astrophysics? Might we also in generations to come clone centenarians, as Haldane proposed, not because a long life is necessarily desirable, but rather because of the invaluable medical insights into aging which would likely come from such a venture?

If asexual cloning becomes the favored mode of reproduction, what would this imply for the traditional sexual mode of reproduction, even as modified by artificial insemination, artificial wombs, and surrogate mothers? If cloning becomes prevalent in a society, sexuality would be totally and completely divorced from *any* reproductive function. In terms of reproduction, human sexuality then could not even look to a contribution to the local sperm or egg bank for consolation! Men may now feel their manliness threatened by artificial insemination. But how will women react emotionally and psychologically to being deprived of their role in reproduction not just as provider of the egg and a nine-month incubator, but also of anything to do with the process at all, from start to finish? A single male, alone on another planet, could theoretically reproduce a whole population from a piece of his own skin, given efficient incubators and a cloning technique!

Haldane's many provocative and controversial suggestions for a cloned population contain some comments on educational policies which might (or might not) be desirable in such a society. Noting that many geniuses have had very unhappy and frustrating childhoods, he proposes that each genius should be responsible for the education of his own clone. This would allow the offspring to benefit directly from the brilliance and skill of their "twin parent," thereby reinforcing their capacities and giving them so to speak a head start. The enriched environment supplied by the parent would then complement the enriched heredity already supplied by cloning. But do geniuses necessarily, or even often, make exceptional or even good teachers? Are geniuses any better, or more stimulating as parents, than ordinary understanding sincere parents? How much of a creative role do childhood frustrations and conflicts play in the flowering of a genius? How important is a good stimulating, enriched environment in the making of a genius? These are not new questions, but psychologists, sociologists, and educators will have to explore them in greater depth before we can appraise the possible values of cloning for intellectual qualities.

Setting aside for the moment the social and personal problems likely to be created by cloning, there are some distinct advantages and disadvantages apparent from a strictly biological and evolutionary vantage point.

In terms of advantages, cloning would eliminate the phenomenon of immune rejection which now plagues practically all skin and organ transplants. Grafting tissues or whole organs between individuals within a clone should pose only minor problems. This ease of organ replacement might become vital to a small group of space explorers on another planet or on a long voyage. A second advantage might be found in an extension of a phenomenon already observed and studied in the psychology and behavior of identical twins. Persons who develop by a splitting of a single fertilized egg, identical twins, very often exhibit a peculiar sympathetic awareness of each other's psyche. Even when raised in totally different situations where they have not been in contact for many years, identical twins often end up marrying the same type of women within a few months of each other, working at similar professions, and expressing personality traits that are remarkably alike. Oftentimes identical twins claim their psychic sympathy amounts to something akin to thought transfer. However, in 1958 Dr. J. B. Rhine, a leading student of extrasensory perception, reported:

> Nothing outstanding has occurred in any single case of identical twins tested so far. The averages on the extrasensory tests were approximately the same whether the sender and receiver were identical twins, fraternal twins, singleton siblings, or simply friends.

Nevertheless, some scientists have argued that the commonly accepted ability of identical twins to understand each other intuitively is one advantage we can expect to find in greater degree within a cloned population. This ability, they go on to point out, might prove very useful in team projects where men must work together under very tense circumstances and for long periods of time. Such applications might come in the underwater surveyor teams of Jacques Cousteau and the SEALAB project off the California coast. Astronauts might also profit by an increase in intuition and psychic communication. On the other hand, one can easily argue that too much similarity and sympathetic intuition might prove a distinct disadvantage in such situations. When lives and the success of an expedition depend on the ability to meet emergencies, perhaps a variety of mental abilities, capacities, reactions, and intuitions rooted in quite different genetic and environmental backgrounds might prove more fruitful than uniform genius. Distinct mental differences among the team members may well prove more efficient and adaptive than cloned uniformity when unexpected crises arise.

Also on the negative side the biologist can argue that cloning reduces variations within the genetic pool of man. Evolution through natural selection depends on variety within the population as its raw material. When the environment changes a population will die out if its individual members are all so well adapted to success in the old environment that there are no mutant forms around to survive the demands of the new climate. A species can only adapt and survive in a new world if there are enough neutral or even harmful genetic variations present in the population as the environment changes. On the other hand, anthropologists are quick to point out that man is no longer the involuntary subject of an uncontrolled environment. Man can and does create his own environment, building houses and air-conditioning them or fitting them out with central heating and electric lighting. Perhaps genetic variety is then no longer vital or even important for man's survival in this world.

Even if we grant this point, there is another aspect of cloning which must be considered. Animal and plant breeders have learned by hard experience that breeding for certain traits carries with it certain definite dangers. Inbreeding can indeed increase such inherited traits as milk production in cows and large breasts in turkeys, but with constant inbreeding for these traits you also increase the chances of harmful recessive traits appearing in the same individuals. Ordinarily these rare recessive traits would be masked by the more common dominant and normal genes, but when selective inbreeding increases the frequency of genes controlling good butter fat production it also increases the frequency of other genes on the same chromosome. Geneticists have long been aware of the value of occasional outbreeding in even the most selected strains. Often when two different strains of animals are crossed the result is an exceptional offspring, possessing valuable traits unknown to either parent: heterosis or hybrid vigor. This being the case, scientists frequently use a combination of these breeding methods, inbreeding within the group as well as crossbreeding with other strains. Perhaps man will have to follow suit and alternate asexual cloning with an occasional "interracial" marriage and the traditional old-fashioned way of procreation.

The virgin birth of human beings and the cloning of people goes far beyond Spemann's 1938 description of the "somewhat fantastical." Yet if there is one lesson we should have learned by this point in our tour of reproductive technology, it is that the *realistic* person does not reject certain possibilities simply because they *sound so incredible*. The impossible, as we have seen several times already, contains an uncanny propensity for becoming a reality in tomorrow's world. What for us today is unbelievable may well be taken for granted by our children.

Stanford University's Nobelist Joshua Lederberg is prone to delectable and sensational off-the-cuff remarks, yet his scientific judgment is more than astute. Consider seriously then his statement that "there is nothing to suggest any particular difficulty about accomplishing 'cloning' in mammals, or man, though it will rightly be admired as a technical *tour de force* when it is first accomplished." The eventual reality of human clones is taken for granted! Lord Louis Rothschild echoes this warning. For years Rothschild has been one of the world's great physiologists, a professor at Oxford University, and an internationally acclaimed expert on sperm activity. He compliments this biological background with the practicalities of the business world, where he has served in a position of some authority in one of the world's largest chemical firms. From this solid base, Rothschild spoke bluntly to scientists at the Weismann Institute of Science in Israel in 1967. At that time he stated rather matter-of-factly that he regards *the cloning of humans as a near possibility soon to be realized.* Obviously then it would be unwise for us to dismiss the subject as ridiculous. To invest some time and effort *now* in exploring the possible implications of cloning might prove to be one of our wisest and most productive endeavors in terms of preparing for man's future.

FIVE

PRENATAL
MONITORING
Manipulating
the Unborn

The women relaxing in comfortable lounging chairs are enclosed
from the shoulders down in plastic tents supported by barrel-
shaped fiberglass frames. To the casual observer they might appear
to be devotees of some curious new hybrid of a beauty parlor
and a weight watchers' reducing salon. But these women are
looking for neither beauty nor slimness. Their purpose is much
more serious as they visit this shrine of the new embryology: a
relatively painless childbirth. At least once a week and often
twice daily for half an hour, starting as early as their eighteenth
week, these expectant mothers visit London's Decompression
Clinic.

Opened in December of 1967, the Clinic is making a practical
application of a controversial technique developed a decade earlier
by Professor Ockert S. Heyns, dean of the Medical Faculty at the
University of Witwatersrand and chief of obstetrics at Queen
Victoria Hospital in Johannesburg, South Africa. A gynecologist
by profession, Dr. Heyns has been deeply concerned with the

complications associated with the latter stages of pregnancy and childbirth. Prime among these is toxemia and eclampsia. Toxemia is a poorly understood condition in which poisonous waste products build up to dangerous levels in the blood and trigger convulsive attacks in the fetus. Toxemia during pregnancy seems to affect certain families, certain races, and even certain areas of a country more than others. Possible explanations for this variation in its occurrence may likely include inherent racial characteristics and dietary habits. In South Africa the incidence of toxemia during pregnancy varies from 17 percent among Cape Malay women and 12 percent among Cape Colored women to between 5 and 8 percent for white and nonwhite expectant mothers in Johannesburg's Queen Victoria and Baragwanath hospitals.

As a pregnancy advances, and particularly during labor contractions, a woman's uterus pushes forward and changes in shape from oval to near spherical. The tensed muscles of the abdomen often interfere with this transformation, causing pain and prolonging the birth. The rather general assumption has been that toxemia and the convulsions it brings on are due to the increasing pressures of the abdominal muscles reducing the flow of blood in the uterus and placenta. If this is the real cause of both toxemia and prolonged labor in many cases, then, Heyns argues, reducing the exterior pressure on the abdomen may allow the muscles to relax and expand, thus allowing the womb also to expand more normally. The end result, hopefully, would be an increased supply of blood to the womb, placenta, and fetus.

With his colleagues Heyns put together a rather simple decompression chamber in which the expectant mother could be treated. Covering the whole trunk of the body this suit had some disadvantages: perspiration, amniotic fluid, and vaginal secretions were retained within the suit and it was uncomfortable. A simpler and more sophisticated "bubble" has been devised by Dr. Louis Quinn of St. Mary's Hospital in Montreal, Canada. Enclosing only the abdomen, Dr. Quinn's patented "Birthezz" device comes in three sizes, costing $290 each or $430 for the set of three. A plastic chamber of some sort and an ordinary vacuum pump capable of reducing the atmospheric pressure inside the bubble by about 20 percent is all this technique requires in the way of equipment. Even with his early apparatus Heyns's results were striking: a dramatic shortening of the labor period and reduced discomfort. The women went into the final stages of labor in a much more relaxed condition and with more reserve strength intact. But more important was the fact that only two out of the three hundred women treated for short-term decompression experienced even mild symptoms of toxemia and pre-eclampsia. With these particular patients 7.6 percent normally would have experienced problems; as it was, less than 1 percent were affected.

Promising as these preliminary results appeared, a more structured scientific test with a control group for comparisons had to be pursued before any real evaluation could be risked. An early test involved one hundred expectant mothers, each suffering from toxemia brought on either by essential hypertension or by a chronic kidney infection. In some cases the toxemia had reached the verge of convulsions, being judged by the doctors as severe or at least moderate pre-eclampsia. Fifty of the women were treated at Queen Victoria Hospital with abdominal decompression, bed rest, and a low-sodium diet. The control group, the other fifty women, were treated at Baragwanath Hospital with bed rest, a low-sodium diet, and the customary drug therapy then in use. On the first day the women in the test group were decompressed for ten minutes in the midmorning and midafternoon. The twice daily decompressions were increased to twenty minutes on the second day and to half an hour on subsequent days. Where feasible, it was even used during labor. During the decompression treatment, normal atmospheric pressures were alternated with the reduced pressure so that the women experienced decompression for fifteen seconds followed by normal pressure for fifteen seconds before another decompression began.

While patients suffering from simple kidney infection and high blood pressure responded about the same to decompression as they did to the drug treatment, twice as many pre-eclampsia patients responded favorably to decompression as to the drug treatment. More than twice as many babies were stillborn or died shortly after birth in the drug-treated control group as in the decompressed group. With the pre-eclamptic mothers decompression reduced infant mortality more than two-thirds over that achieved with drugs. These are very promising results. On the other hand, decompression is far from being a panacea for all cases of prenatal toxemia, hypertension, kidney infection, and eclampsia. Cure-alls simply do not exist in the real world of medicine and a treatment that helps one patient may not benefit another at all.

I have gone into some medical detail on this decompression technique for two reasons. First, all but two of the semipopular and popular discussions I have seen of Dr. Heyns's technique fail to mention anything about the solid medical background which is the real substance of his work. For the sake of sensational enticement, most who have commented on Heyns's work prefer to pass over the medical aspects with a brief one-sentence nod smothered somewhere in a melodramatic account of a psychological spin-off of this work. Yet without this background, the whole story becomes unbalanced and distorted. At the same time, the reader tends to miss a very important aspect of all scientific endeavors. This is my second reason for presenting the medical background. Very often a scientist sets off as Dr. Heyns did to solve a particular problem, in this case the problem of toxemia and prenatal eclampsia. This is the only goal in view, yet the unexpected side effects of an experiment can on occasion become even more important than the original purpose. A new technique is developed with a particular application in mind, and other applications often pop up unexpectedly. When Dr. Ross Granville Harrison pioneered in the *in vitro* culturing of tissues, he had no idea he was laying the groundwork that made possible the discovery of the Salk polio vaccine. Likewise, the doctors at Eli Lilly Pharmaceuticals had no idea that their testing of an extract from the periwinkle flower which natives in South America use to cure everything from scurvy to dysentery would produce a drug helpful in the treatment of certain cancers. Science is exploration and the outcome is more often than not unpredictable.

Dr. Ockert Heyns's decompression work is a perfect example of the unexpected spin-off of scientific research. The side effect first came to light when some of the mothers in Heyns's original groups called to tell the doctor about their wonderful children. They all seemed to be extraordinarily bright and gifted. This is a natural reaction for most new parents, especially with their first-born, and Dr. Heyns nodded knowingly as he chalked up their remarks to human nature and parental pride. But the reports kept coming in. Before long Dr. Heyns began to wonder if there might not be some substance to them. At the age of thirteen months, little Katl Oertel was dialing and answering the telephone. By the age of three Katl was speaking in four languages. Children in South Africa often hear English, German, Zulu, and Afrikaans spoken around the house, but normally babies only pick up the language their mothers speak.

The decompressed babies were bored at their nursery schools and chatted with adults in a fluent matter-of-fact way quite unusual in young children. While the average child has a vocabulary of perhaps half a dozen words by the age of eighteen months, decompressed children were commonly handling two hundred words. On the Arnold Gesell test for infant development decompressed white South African children scored about 18 percent higher on the average than nondecompressed babies. In one such test group, 16 percent of the decompressed babies scored over 48 percent higher. On their first birthdays, six babies who had been watched rather closely after experiencing decompression during gestation and birth appeared to match the physical and mental developments of a normal two-year-old.

The explanation Heyns offers for this psychological effect does have some weight of medical logic behind it. In the final weeks of a pregnancy, the human placenta practically stops growing though the embryo keeps developing. The infant's tiny heart must then meet the increasing task of pushing blood at ever higher volumes through the placenta. Unequal to the task, the fetus does not receive as much blood and oxygen as it could use and often needs merely to survive. The result, Heyns believes, is that the brains of most infants fail to develop to their full prenatal capacity for lack of the necessary oxygen. By reducing the atmospheric pressure, even for short periods of time, Heyns believes the oxygen supply through the placenta is increased enough to account for the improved psychological and physical developments.

It is interesting that while Heyns and other researchers have published considerable data on the medical effects of decompression, very little detail has been published on the psychological effects of superoxygenating babies in the womb. An extensive study, covering four years, was undertaken by Dr. Renée Liddicoat at Queen Victoria Hospital with Dr. Heyns's cooperation. Expectant mothers, selected at random, were divided into two groups; half were given decompression and the other half physiotherapy. The babies were tested at four weeks, and at four and nine months with the South African Child Development Scale and at the age of three with the Merrill-Palmer test for I.Q. which Dr. Liddicoat maintains is more reliable than the Gesell test used by Heyns. At the age of nine months decompressed babies appeared to have a slight edge, but by the age of three the two groups showed no significant difference in intelligence. Dr. Liddicoat suggests that the "superbabies" observed by Heyns were likely due to a combination of the probable higher intelligence of the women who requested decompression and accidental biasing in the administration of the Gesell tests in Heyns's study. Dr. Heyns, on the other hand, contends that Liddicoat's experiment did not give the patients adequate exposure to the decompression bubble. Despite this inadequacy, Heyns's own I.Q. tests on 101 of Dr. Liddicoat's superoxygenated babies showed significant intelligence differences in the nine- to thirteen-month age range. Two psychological studies, each reaching the opposite conclusion, and as yet no impartial study to resolve the confusion. Disturbed by the sensationalist press coverage of sixteen decompression clinics which opened in England in the winter of 1967–1968, the *British Medical Journal* sided with Dr. Liddicoat's findings, but urged that more rigorous and thorough experiments be conducted *soon* "before the trickle of mothers becomes a flood, perhaps with disappointments for them and unhappy consequences for their children" when mothers expect too much of their decompressed babies.

It will be some years before any real evaluation can be made of the psychological side effects of superoxygenation. These long-range effects are in reality the crucial ones, and some basic questions about the technique lie in that area. If we grant that decompressed babies do indeed receive a head start, will this have a lasting effect on their development? Or will they simply be "fast starters" who then fall back into the crowd in childhood or adolescence? Some indications point in that direction. But if the head start does have a lasting effect, how is this likely to affect the overall personality development? We are already well aware of the emotional and psychological adjustment problems and competition faced by exceptionally bright youngsters who enter college at ten or twelve.

Medically the decompression treatment is in a similar bind of confusion, though our data is somewhat less sparse. In the twelve years between 1955 and 1967 over five thousand women have had babies with decompression treatments at six South African clinics. Some British doctors have become very interested in the idea, and three dozen national health hospitals have installed the necessary equipment despite the fact that the cost of this treatment is not defrayed by the British National Health Service insurance. Even the prestigious University of London Hospital has added decompression equipment to its facilities.

In Guadalajara, Mexico, Dr. Moises Hernandez Gutierrez operates a maternity hospital which in effect is a decompression clinic. He agrees with Heyns that decompressed babies are brighter than average and at birth have a better coloration, reflexes, and respiration. He also observed a significant reduction in labor. However, these are only observations, and the statistics have not been forthcoming. At St. Mary's Hospital in Montreal, Canada, Dr. Louis J. Quinn has reported his experience with 593 decompression births over a period of four years. Writing in the *Journal of Obstetrics and Gynecology,* Dr. Quinn claimed that the technique shortened the first stage of labor so that 83 percent of his patients delivered within five hours. He also reported a reduction in the pains of labor and an above average Apgar score for the decompressed infants. (The Apgar Score is a 0 to 10 scale rating of an infant's chances for survival based on skin color, breathing, heartbeat, and reflexes immediately after birth.)

Few American physicians have explored the effects of decompression and few of these are favorable to it. Among the thirty or so American doctors to try decompression is Daniel O'Keefe, an obstetrician in Glens Falls, New York. After nine years of using the bubble in the first stage of labor, Dr. O'Keefe has found that it lessens the pain and reduces the need for medication. The only drawback is that its use requires explanation and monitoring, a time-consuming expense for any physician. Counter to this has been the experience of two New York City obstetricians, Drs. Howard Shulman and Stanley J. Birnbau, who tried the bubble on twenty-five selected patients. "Little or no effect on the duration of labor . . . little or no reduction in labor pain . . . no ill effects," were their conclusions from this study which has been faulted as too limited in its sampling. Much more extensive has been the experience of the "bubble's" staunchest and most vocal American advocate, Dr. Lawrence E. Lundgren. A fifty-one-year-old Texan and an inactive affiliate of the Baylor College of Medicine, Dr. Lundgren has concentrated on a private practice in obstetrics. A "bubble" convert since 1963, Dr. Lundgren will not accept expectant mothers unless they agree to use the Birthezz "bubble" of which he has twenty-three in constant home use by patients during the late stages of pregnancy. As with Heyns and others, Lundgren maintains that decompression shortens the labor period and pains and produces babies with above average Apgar scores. These conclusions came in Lundgren's report of his first 412 cases of decompression to the Texas Medical Society convention in 1969. In 1968 Lundgren authored a section on decompression for the standard medical text, Davis' *Gynecology and Obstetrics.*

Generally speaking, the American reaction has been one of indifference and skepticism. Spurred apparently by Lundgren's work, doctors at Baylor tried a pilot test on three dozen expectant mothers. With negative results on this small sample, the study was not pursued, and Dr. Lundgren's claims were chalked up to his personal care of patients inducing a more relaxed delivery than service patients handled in the Baylor clinic. The American College of Obstetricians and Gynecologists has expressed no official opinion about the technique. The American Medical Association has, however, refused to allow exhibition of the Birthezz device at its annual convention because of insufficient documentation of its effects. Other major medical conventions, on the other hand, have allowed the "baby bubble's" manufacturers to exhibit their device. Interest expressed at the Yerkes Primate Center in Atlanta may well result in a full and solid experimental testing of the decompression technique.

The medical questions, and even to some extent the immediate psychological effects of the decompression technique are quite straightforward and uncomplicated. More subtle, however, are its implications and long-range effects. All the basic questions raised by man's new ability to directly modify the process of pregnancy and birth come to focus in discussing the outcome of decompression for man's future. Fetology, the new science which is grappling with these implications, is at most ten or fifteen years old. Its field is monitoring the life of the unborn, and then from the information so gained modifying the fetus in the womb either to correct an abnormal condition or perhaps improve on a normal situation. In fetology we add a whole new context of overtones to our queries about human values and ethics. In monitoring and modifying the fetus in the womb we are not concerned except in a very peripheral way with the parents or society in general. Our prime interest is performing for the unborn certain often extraordinary services which will enrich his life as an adult human being. With fetology our intent is the future and the uneasy foresight of a seer is often our only guide.

During the intrauterine life of the fetus there is, according to Jean Rostand, a set number of cell divisions. When the thirty-third general division has occurred the infant is ready to enter this world and the labor contractions begin. Since the biologist is now uncovering many drugs which will inhibit or stimulate cell division, Rostand wonders if perhaps we should search for a drug which will selectively trigger cell division in brain cells only. If such a drug were found and used with the unborn, we might double the size of its brain before birth and produce the superman foreseen by Renan in 1871, a human of a superior race whose highly developed nervous system would greatly increase his powers of reason and perception. Of course, Caesarean section or an artificial womb would then be required. Many biologists are skeptical about the possibilities of such differential increases in the human brain, yet it is something worth considering at least as a possibility (or threat).

A crucial question here is the relationship between brain size and intellectual capacity. At birth the human baby has achieved only 23 percent of its ultimate brain volume compared with more than 50 percent at birth for an ape. Only around its first birthday does the human baby catch up with the baby ape in the percentage of adult development achieved. A comparison of the average brain capacity in different fossil and contemporary humans can add some lively fuel to the question. The cranial cavities of the very early Australopithecines of South Africa, "near-men" in the mind of most anthropologists, average between 450 and 550 cubic centimeters. The Java men average 770 to 1,000 and their cousins, the men and women of Peking half a million years ago, average 900 to 1,200 cubic centimeters. Neanderthal man, a very complex group with many forms, ranged roughly between 1,300 and 1,425, while recent modern-type man ranges between 1,200 and 1,500. In every group, however, there are always the exceptions and extremes. Anatole France, whose genius few doubt, was a small man with a small brain, only about 1,000 cc. while Ivan Turgenev, an equally famous author, was a big man with a brain capacity almost twice that of France, 2,012 cc.! Paleontologists M. Boule and H. V. Vallois gave some other figures for brain capacity in their summary of 1957: 1,965 cc. for Bismarck, 1,830 cc. for Baron Cuvier, the founder of comparative anatomy. Adult males in Paris average 1,340 cc., in Auvergne, 1,585 and in the Australian bush country, 1,300.

Geneticist Theodosius Dobzhansky offers a terse comment on this in his work *Mankind Evolving:*

> Large cranial capacity is evidently not indispensable for high intelligence or achievement. But it is a fallacy to conclude that since brain size alone does not unalterably set the level of intelligence, the two variables are not in any way related. Such a conclusion probably again reflects the misconception that a trait is either wholly genetic or wholly environmental.

Dobzhansky goes on to summarize the work of B. Rensch and his co-workers comparing the brain size with long-term memory, the speed of learning, and the capacity for learning. Using pairs and groups of related species and races with contrasting body and brain sizes—white rats and mice, elephants and horses, donkeys and zebras, large, medium, and small chickens, large and small fish—Rensch has found that the speed of learning cannot be directly correlated with the size of the brain, though the average number of "tasks" learned depends very much on the size of the brain. Memory is definitely superior in the larger animals.

The "best" rat remembered all his learned tasks for 154 days; the "best" mouse, for only 103 days; the elephant, for over a year, and a horse for six months. Rensch concluded that the memory retention of the animals experimented with by himself and others is about proportional to the brain size. Perhaps, he cautiously suggests, the larger brains contain nerve cells with more numerous branches or dendrites which permit many more interconnections between the cells and hence more pathways for learning and memory in the brain.

The simple gross volume of the brain is not nearly as important as the complexity of the cerebral cortex where reasoning occurs and the ratio of body/brain size. The elephant's brain is at least three times the size of a human brain but most of its activities are consumed in merely coordinating the functions of its immense body.

Even so, it poses an interesting dilemma for man tomorrow to ask whether we should, if we could, modify the brain development of the fetus.

Some steps have already been taken in this direction. On a rather crude plane, two Swedish experimenters, Haggquist and Bane, have done some interesting work with triploid rabbits which turn out to be *precocious giants*. The normal animal has a diploid complement of chromosomes in each cell, two complete paired sets or $2n$ in scientific shorthand. In some plants and animals it is possible to treat the reproductive glands with a chemical like colchicine so that the developing gametes end up with twice the usual single set of chromosomes; they are then $2n$ instead of $1n$. When two normal gametes, an egg and a sperm each with one complete set of chromsomes ($1n$), get together the fertilization results in a normal $2n$ embryo. But when a sperm or egg has $2n$ to begin with, fertilization results in an animal or plant with three sets of paired chromosomes in each body cell, $3n$. The Japanese have produced a triploid watermelon which has the delightful trait of being seedless. Haggquist and Bane's triploid rabbits, according to Rostand, are abnormally large and quite precocious. What about the possibilities, then, of using colchicine or some similar drug on a human ovary or testes, *in vivo* or *in vitro*, to produce diploid eggs or sperm and thus a triploid human?

A simple increase in the size of the human brain is a rather primitive and crude way to improve intellectual capacities when one considers the sophistication of modern biochemistry and physiological manipulation. Dr. S. Zamenhof and co-workers at the University of California have injected pregnant mice and rats with pituitary growth hormone while the fetal brains were still maturing, from the seventh to the twelfth day of pregnancy. When the offspring were later sacrificed and examined, not only were the brains much heavier than normal, but there was also a definite increase in the number of neurons in comparison with the supporting glia cells. More important though was the discovery that the neurons in the cortex were much denser than usual and had many more branchings connecting the cells. Another research team, working in the same area, has suggested that these experimental animals have perhaps 50 percent more contacts between neurons than normal. This increase in both density and contacts one would expect to result in better learning and a higher intelligence in the experimental animals. In standard maze tests the manipulated rats were definitely better in performance than the normal animals. Another animal whose brain has been manipulated for an increase in intelligence is the tadpole.

Paralleling the theory of Dr. Heyns is the suggestion of Dr. Huidzie that blood circulation and the complexity of the blood vessels nourishing the brain may be an important factor influencing our intelligence. From another angle chemical stimulation of the brain, CSB as Albert Rosenfeld terms it in *Second Genesis,* may prove practical in the fetology of tomorrow. Cholinesterase, an enzyme which acts on acetylcholine in the nervous system cycle, may provide a clue to improving our basic intelligence. Several researchers, among them David Krech and Edward L. Bennett, have found this enzyme more prevalent in the brains of more intelligent rats. Other chemicals, such as glutamic acid, help mice solve maze problems much faster than normal. If occasionally we use such chemicals in treating certain cases of mental retardation, what about the possibility of using them or similar drugs to give a human fetus a head start in the learning race? Even simple ions of potassium, injected directly into the brain by Dr. Roy John, enhance the learning process while calcium ions retard it very strikingly. This whole area of research is in *a very primitive state today*, but does that fact make it any too early to start meditating on the possible applications and implications?

The key to fetology is man's growing ability to monitor the development of the embryo from fertilization to birth. The private world of the womb can now be probed by a dozen different devices which extend and complement the inquisitive eye and ear of man. Many people are still rightly fascinated by the color photographs of the embryo in its amniotic sac which appeared a while back in *Life* magazine. The incredible detail, lovely color, and almost eerie lighting of these portraits are unforgettable. Yet they represent only a very early step in man's probing of life in the womb. Color television broadcasts from inside the human body are now a reality. Pioneered by Dr. Irving Bush, head of the urology department at Cook County Hospital in Chicago, the first subject of a television broadcast from inside a human body was a view of the lining of the urinary bladder. The bladder was chosen for this premiere because it is the most difficult internal organ to reach.

Dr. Bush modified an ordinary cystoscope by adding a fiber optics light pipe to the mirrors and tube of the cystoscope and then linking this on the outside to a television camera which projects the internal scene on a large viewing screen. The color broadcast can also be recorded on video tape for later replay and study.

At London's St. Mary's Hospital doctors are using a device similar to that developed by Dr. Bush to check on the color of the amniotic fluid which surrounds the baby in the womb. If the fluid is yellow or greenish instead of the normal milky white, the doctors know that something is going wrong in the pregnancy. They can also check with this technique, known as amnioscopy, to make sure the right amount of amniotic fluid surrounds the embryo's head.

What goes on in the dark moist world of the womb? In the past our imagination has pictured the tiny embryo resting peacefully and passively waiting for the time of "quickening" to arrive five months after conception so it can let its mother know of its presence. Intrauterine television has shown us a quite different

picture. Long before quickening the fetus is exercising its muscles and learning little skills which will come in very handy in
later life. Dr. Margaret Liley, of the National Women's Hospital
in Auckland, New Zealand, has described the young fetus as an
"unborn human astronaut," because of its careless free activities
in the watery world of the amniotic sac. Weightless, buoyant,
and as active as an astronaut on a spacewalk, the fetus is swimming at the age of three months, exercising its arms and legs,
reacting to sounds, eating and feeling pain though it weighs
scarcely an ounce and is only three inches long.

The husband and wife team of Margaret and William Liley
have been interested in the life of the unborn for years. Together
they have used intrauterine photography and X-ray fluoroscopy
with closed-circuit television to study the fetus in his natural en-
viron. In the early months of pregnancy when space in the womb
is not at a premium, the fetus practices some acrobatic endeavors,
often somersaulting twice in as many minutes. Later, as space
becomes more cramped, he is primarily concerned with finding a
comfortable position as his mother moves around during the day.
At times, when his mother is standing, this means wiggling into
a hammock-style sling position in the womb. At other times,
when the mother is lying down, such a position may bring the
child back up against its mother's bony spinal column and a shift
to a side position may be in order. All in all, according to the
Lileys, the embryo is very active, gaining the muscular abilities
on which his life will soon depend.

Inside the womb the fetus learns to eat by drinking the amniotic fluid. About the fourteenth week, Dr. Margaret Liley reports,
the fetus is "practicing the arts of sucking and swallowing which
will insure his survival when born." This is no small matter, for
an active fetus will consume some six to eight pints of amniotic
fluid a day, a real test for his developing digestive system and
kidneys. Though the embryo receives most of its food from diffusion across the placenta, analysis of the amniotic fluid indicates
that it is a fair source of calories, containing small amounts of
both proteins and sugars. In terms of actual nourishment the six
to eight pints of amniotic fluid is the equivalent of about three
and a half ounces of milk. Commenting on this, Dr. Liley suggests
that infants with a narrow or constricted esophagus are apt to be

smaller at birth because they cannot drink very well while in the womb.

Even the world of sound and music is open to the unborn, for the Lileys have discovered that the embryo actually hears sounds while seemingly cut off from the outside world. The fetus has been observed reacting to its mother's heartbeat and sharp noises like the backfire from an automobile. The effervescence of Schwepp's Bitter Lemon, a goblet of champagne, or a mug of beer reach the fetus with a rumble "akin to rockets being shot off."

Ancient superstitions and old wives' tales have always warned expectant mothers against emotional outbursts and anger, for these would "mark the child for life." The new evidence that the fetus does indeed react to sounds and its mother's movements might at first sight seem to substantiate that belief, but Dr. Liley is quick to dispel the connection. Sounds for an unborn baby are without emotional impact, for he has nothing in his experience to which he can relate the sound unpleasantly. An outburst of uncontrolled anger may trigger an emotional response in a young child of two or three when he is the brunt of physical abuse, but not so with the unborn. "It is ridiculous and unhealthy to assume that a woman must restrain her emotions because she is carrying a baby," Dr. Liley writes in her description of *Modern Motherhood*. "When an expectant mother laughs, rages, weeps or shouts, it has about the same effect on her unborn baby as if she had dropped a saucepan."

Fetology, in the words of Dr. Sheldon Cherry, a pioneer in the field at New York's Mount Sinai Hospital, "has humanized the fetus."

The visual coverage of gestation is today complemented by a variety of other scientific monitors, some of limited or passing practicality and others promising for extensive use in the years to come. Thermography, for instance, can be used in certain diagnostic situations. This involves a heat-sensitive photographic plate which turns the heat radiating from the mother and fetus into a visual image. Very similar to the infrared night photography equipment used by the army and spy satellites, this technique depends on the different amounts of heat radiating from different parts of the mother's body. Normally the surface heat of our body is fairly even so that an infrared photograph shows only an outline with no contrast or detail. However, when a woman is pregnant, her breasts and parts of the abdomen are hotter than usual. This often provides a fair picture of the placenta and warns the doctor of such abnormal and dangerous developments as internal bleeding or a hemorrhage in the placenta. Infrared photography is limited, however, to surface outlines and cannot probe very deeply. Hence the doctors occasionally resort to ultrasonics, or very high-frequency sound waves, another technique still in the research stage. High-frequency sound waves can probe beneath the surface and provide a picture detailing even the soft parts of the infant's body, something impossible with the X-ray image which reveals only bones and a faint body outline. Ultrasonics can detect abnormalities such as a hydrocephalic condition where excess cerebral fluid dangerously swells the fetus's head so that a normal birth is impossible and mental retardation an inevitability.

One of the major dangers a human fetus must risk in entering this world is the possibility that, as he squeezes through the narrow birth canal, defects in the placenta or compression of the umbilical cord may reduce his supply of oxygen and nourishment. Fetal distress, as this condition is known, accounts for many, if not the majority of some 28,000 apparently normal babies who die annually in the United States in the process of being born. Until recently the physician delivering a baby could rely only on indirect indications of this complication. Usually the obstetrician or an assisting nurse tries to detect possible trouble by listening to the fetal heartbeat with a stethoscope. As long as the rate remains within the normal range of between 120 and 160 beats per minute, the assumption is that all is well. However, fetal distress is most likely to occur during the contractions of the womb when muscular movements prevent the doctor or nurse from hearing the heartbeat. In between contractions a baby's heartbeat may fall within the normal range even though he is suffering from severe oxygen shortages during the periodic contractions.

Medically the stethoscope is an antique, though still a generally useful instrument. In recent years its role in labor and birth has been complemented and even replaced with far better results by electronic monitors for the unborn. Drs. Mortimer Rosen and Joseph Scibetta are using one such monitor in their regular practice at the University of Rochester's Strong Memorial Hospital. As soon as the cervix, the opening of the womb to the birth canal, has dilated, doctors can use an endoscope—a hollow tube with a viewing light at one end—to attach silver electrodes to the head, buttocks or a handy limb of the fetus as it begins its journey through the vagina. When an obstetrician anticipates possible complications he may rely on such an electrode to monitor the infant's brain waves. Information such as this may not only aid in the delivery of a normal child, it may also supply helpful insights into the causes of brain damage during childbirth.

At the University of Washington, Dr. Wayne Johnson has developed another type of monitor capable of detecting the fetal heartbeat only twelve weeks after conception. At hospitals in Rochester and Chicago doctors are experimenting with thermistors and other monitors during labor, attaching them to the scalp as the baby's head first appears at the upper end of the birth canal.

The first medical clinic in the world to put into operation, as a regular practice, an intensive-care prenatal unit for electronically monitoring fetuses from the onset of labor through to birth opened in the fall of 1968 at the Yale-New Haven Hospital in Connecticut. Dr. Edward H. Hon, its originator and director, is one of a new breed in medicine, a specialist trained in both obstetrics and electronics as a biomedical engineer. A Chinese-born professor at Yale University School of Medicine, Dr. Hon has spent the last thirteen years perfecting an electronic monitoring system which will not only provide a continuous record of the fetal heartbeat but also continual information about the onset, rate, and pressure of uterine contractions. Readings from these instruments are recorded simultaneously on graph paper by two monitor panels, one beside the mother in the labor room and the other at a central nursing station nearby.

By correlating the labor contractions and the heart rate changes, Dr. Hon's continuous monitoring system has shown that a simple reliance on changes in the heartbeat can be very misleading. Actually there are three possible major changes in that rate and only two of these are potentially dangerous for the infant.

The first change in the pulse comes during the compression of the head as it passes through the cervix and pelvic girdle. The soft, somewhat malleable skull of the baby is squeezed, with the resulting pressure causing a slight drop in the heart rate which then returns to normal as the contraction passes. This steady drop is perfectly normal and harmless to the infant.

A second, potentially dangerous drop in the heart rate comes if there is an insufficient supply of blood to the placenta. The signal here, a drop after a contraction has begun, warns the doctor to reduce the amount of anesthetic being used or the dosage of drugs used to induce labor. He might order oxygen for the mother, or in a last resort deliver the child by Caesarean section. Nine out of ten cases of fetal distress are brought on by constriction of the umbilical cord. With this the rate of the heartbeat often becomes wildly erratic with no seeming correlation to the labor contractions. An obstetrician relying on the stethoscope alone might consider such a rapid drop in pulse and the erratic rate serious enough to order a Caesarean section. However, electronic monitoring has shown that most cases of cord compression can be remedied without surgery simply by shifting the mother's position. In fact, with continuous prenatal monitoring, three-quarters of the babies that would in the past have been delivered by a Caesarean section at the Yale Hospital have been delivered normally and without complications. In a test with four hundred mothers known to have histories of difficult deliveries, none of the babies died and injuries were halved.

Monitoring units, each capable of handling four babies at once, are now installed in most of the labor rooms at the Yale Hospital, and eventually all the babies born there will benefit from this new system. Dr. Hon's device will no doubt be introduced, with increasingly new improvement, into other hospitals around the world, further reducing the number of fetal deaths and injuries as well as the number of mentally retarded children born.

Amniocentesis is the technical name given to another relatively recent method of checking on the development of the unborn.

Though it does involve some risk and is not used without serious reason, the procedure has several useful applications. It first came into prominence in conjunction with the Rh blood factor which frequently causes complications for the fetus. This arises when a blood protein, known as the Rh factor for the Rhesus monkey in whose blood it was first found, leaks through the placental barrier from child to mother. If the mother is Rh negative and does not have that specific protein, her body reacts against the strange substance by building antibodies that destroy it. Usually there are no complications in a first pregnancy when the mother is Rh negative and the baby Rh positive. Not until late in pregnancy do red blood cells cross the placenta and then only in small numbers. But at the first birth the hemorrhaging placenta introduces a flow of fetal blood into the mother's system to trigger a buildup of antibodies against the Rh protein. These antibodies remain in the mother's system as a permanent protection, ever ready to attack the foreign protein which a second Rh positive child might bring. In the event of a second such pregnancy maternal antibodies may leak across the placenta and attack the infant's blood very early in pregnancy, bringing blood clots, brain damage, and even death.

For many years after they learned of the Rh factor physicians could only wait helplessly hoping that the second or third child would escape the antibodies long enough to develop to the point where a premature delivery could be safely induced. Hardly out of the womb, the infant was then immediately given a complete blood transfusion to wash out all the maternal antibodies. The hope was that this would all be done soon enough to avoid serious permanent damage to the child, but this was not often the case. Obstetricians soon became more optimistic in dealing with the Rh problem because, in the early 1950's, a British doctor decided to ignore the centuries-old taboo which placed the fetus outside the pale of medicine as long as it remained in the womb. "Let nature be and keep your hands off" was the general dictum violated by Dr. D. C. A. Bevis when he injected a needle through the abdominal wall of an expectant Rh negative mother and withdrew a sample of the amniotic fluid surrounding her baby in the womb. By analyzing the bilirubin pigment in the amber fluid, Dr. Bevis could learn to what extent the infant's blood was being destroyed by the mother's antibodies. If the fetus was suffering from severe anemia and excreting into the amniotic sac large quantities of bilirubin from destroyed red blood cells, a premature labor could be induced and the child given a complete transfusion as soon as it entered this world. This practice, now known as amniocentesis, cut the death rate from the Rh factor by more than 50 percent in many hospitals.

Still, a satisfactory solution to the Rh problem was far from being in hand. The complications of the Rh factor can threaten life as early as the fourth month of gestation, but doctors cannot safely induce premature labor until the seventh month. In 1963 chance brought the physicians another step closer to a solution when Dr. A. William Liley of New Zealand accidentally pierced the abdomen of a fetus while taking an amniotic sample. When the fetus proved no worse for the accidental puncturing, Dr. Liley decided to try transfusing blood into the embryo while it was still within the womb. The fetus proved to have a remarkable and rather unique ability to absorb red blood cells into its circulatory system when these were injected into its abdomen.

The problem then became one of marksmanship: how to aim the needle through the maternal abdomen and into the infant's abdomen without causing any damage. One solution involved the use of a dye which absorbs X rays. Injected into the mother's bloodstream, the dye migrates across the placental barrier into the blood vessels of the fetus, making them visible to the doctor watching the fluoroscope. With one eye on the screen, the doctor can guide his hypodermic needle directly to a large fetal blood vessel. A complete transfusion can be given in this way, often as early as the twenty-second week. Dr. John T. Queenan prefers to have the fetus swallow an X-ray absorbent dye injected into the amniotic fluid. The dye lights up the infant's digestive tract and again the transfusion needle can be guided safely to its target. At the Yale University Hospital, Dr. Maelyn Wade uses a fiber-optic amnioscope in an even simpler and safer approach to intrauterine transfusions. Guided by heatless light from the fiber optics, the fetologist can insert a needlelike instrument either through the abdomen or through the birth canal and into the amniotic sac. There a tiny transparent balloon on the end of the scope inflates to serve as a sort of underwater mask for better viewing of the fetus. Dr. Wade has used this technique with great success, in one case administering five transfusions to one fetus before birth.

This is a risky solution, but when the chances of the mother's protective antibodies attacking and destroying the baby's blood are high, it may be the only way to save the child. Adding to the risk of administering a blood transfusion in the womb is the danger of radiation affecting the fetus. Despite the dangers, intrauterine transfusions quickly became almost standard procedure in hospitals around the world. The death rates in Rh babies dropped even lower, from one in five to one in fifty.

In 1963 the coupling of amniocentesis and prenatal transfusions was a major advance. Today a much safer technique involves immunization of the mother to the Rh protein beforehand. Half a century ago, in 1909, a great American bacteriologist, Dr. Theobald Smith, found that animals develop a passive immunity to a foreign protein when antibodies manufactured by another animal for it are injected into them. The antibody injection seems to paralyze the normal immune response of the host so that it does not manufacture its own antibodies. For decades passive immunity remained a curious but useless fact. Then in the early 1960's an application was found. Drs. Vincent Freda and John Gorman from Columbia-Presbyterian Hospital in Manhattan joined with Dr. William Pollack and others at Ortho Research Foundation in New Jersey to forge a major breakthrough. They found that a serum containing anti-Rh antibodies injected into an Rh negative woman tricks her body into ignoring the presence of foreign Rh proteins which might leak into her system from a child. This passive immunity offers a simple but marvelous protection for the Rh positive fetus.

The new vaccine, known as Rho-GAM, has proven 99 percent effective when administered to a woman within seventy-two hours after the birth of her first child. With each subsequent child the injection must be repeated, for the injected antibodies bring only a temporary paralysis of the immune system. Once an Rh negative mother's system has been activated to produce its own antibodies, however, the vaccine has no effect. With one out of every eight American women lacking the Rh protein and with ten thousand babies dying each year in the United States and another twenty thousand suffering congenital anemia, heart failure, or mental retardation as a result of the Rh complication, this vaccine is a long-sought-for remedy, as welcomed by the physician as by parents.

Closely related to this development is another technique, still very much in the experimental stage. Blood transfusions have raised countless thorny issues in human psychology and medical ethics ever since King Louis XIV's court physician, Jean Baptiste Denis, saved the life of a fifteen-year-old boy who had been bled almost to death for an obscure fever by infusing into his veins nine ounces of lamb blood. When another patient, similarly treated, died, the charge of murder was raised in Paris. Patients who survived the crude operation began to worry about growing hooves and hair while ecclesiastics fumed about the immorality and unnaturalness of the procedure. Not until 1900 did blood transfusions really become popular after the Viennese Karl Landsteiner discovered four distinct blood types, A, B, AB, and O, whose different proteins and antibodies were incompatible.

In 1937 Chicago's Cook County Hospital opened the first American blood bank and transfusions entered the scene as a major medical practice. During the Second World War thousands of lives were saved despite the limitations imposed by the incompatibilities of the four types and the fears of some Americans about the effect of mixing "white" and "black" blood. The "universal donor," a person with O type blood, can give a transfusion to anyone but can only receive blood from another O type person. Most of the time this limitation is a mere inconvenience. The situation became more serious with the discovery of several other protein and antibody differences in the blood (M-N-S, and eight subvariations of the Rh gene) and several very rare types for whom donors are very limited, sometimes to one or two other persons in the United States. A parallel problem has arisen with skin and organ transplants where immunity to a foreign tissue often reduces the chances for success. However, there is a peculiarity about the various immune responses in the human body. They are not always active and in fact develop at different times and rates during early infancy. Thus the immune response to foreign blood comes into action about six months after birth.

Considering this aspect of the immunity question, several research teams have been exploring the possibility of suppressing or paralyzing the immune response before it arises in children with rare types of blood so that later in life they can receive transfusions with more common blood types. A Mexican research team from the National Institute of Nutrition and the Maternidad have joined with Dr. Bernard Pirofsky at the University of Oregon Medical School to test this possibility. They began by injecting newborn O type babies with A type blood and found that the immune reaction could be suppressed for about six weeks before another injection was required. As long as the injections were kept up on a regular basis the baby would not reject A type blood in a transfusion. The Mexican group is now experimenting with immune suppression by feeding foreign blood proteins to the fetus through the amniotic fluid it swallows. The results have been very promising, though the doctors remain equally cautious about the prospects for this experimental, relatively untested technique. However, it does seem to hold possibilities for modifying the fetal immune reaction so that transfusions and perhaps even organ transplants may be much easier for adults in the years to come. A child in the womb, for example, might be physiologically adapted to the tissues and organs of several relatives so that in later life they might serve as donor of an organ if the need arose. In theory, at least, this experimental technique or some modification of it might eliminate the whole problem of immune rejections and make every baby a universal donor and recipient.

At the Regional Primate Research Center in Beaverton, Oregon, several important projects are under way. With a technique similar to Pirofsky's reliance on the fetus drinking amniotic fluid, Dr. Richard Behrman, for instance, is studying the effect of different drugs on the oxygen supply to the fetus. By injecting radioactively labeled drugs into the amniotic fluid, he can learn how fast these reach the embryo, where they locate in the fetus, and perhaps what effect they have. Such studies may give us definite information on the effects of a variety of drugs ranging from nicotine and the tranquilizers to marihuana and LSD when used by expectant mothers.

Semidelivery is a prime monitoring and manipulative technique at the Oregon Primate Center. Thus far it has been successfully tested on lambs, dogs, and monkeys, though only half a dozen human babies have had this experience. The research team which pioneered semidelivery is typical of most in the world of science today. The days of major discoveries by single geniuses like Harvey, Vesalius, Newton, or Galileo have given way to the research team, in the present case bringing together talent from Columbia-Presbyterian Hospital in New York City, the University of Vermont College of Medicine on Lake Champlain and, across the North American continent, from the Beaverton Laboratory. The procedure for semidelivery calls for anesthetizing an expectant animal and thus automatically anesthetizing the fetus in the womb. Then an operation similar to a Caesarean section is begun, the amniotic sac containing the embryo is carefully lifted from the womb, slit open, and the amniotic fluid removed and saved. The fetus can then be observed and examined for up to three hours provided it is kept moist with a physiological saline solution. Its oxygen and nourishment continue to flow from mother to child through the still-attached umbilical cord. A variety of electrodes can be attached to the fetus, to monitor its brain waves, heartbeat, and other physiological reactions. Tiny biologically inert catheters or tubules can be inserted in the umbilical cord, wrist, neck, or leg of the fetus for similar monitoring of the blood for enzyme and hormone levels. After all the desired tests and observations have been made, the fetus is gently returned to its amniotic sac, refloated in fluid, and the incision resewn. Finally the whole package is returned to the sheltering maternal womb, occasionally with monitors still attached to the fetus, and the incision repaired. Despite the fact that the embryo may have spent some hours in a world for which at the moment he is ill-prepared to survive, these fetuses often go on to a normal birth apparently unaffected by their preview.

Semidelivery has been tried on a variety of animals and the information thus gained has carried man several important steps along the road to a better understanding of the nine months we all have spent in the womb. Cigarettes have come in for considerable criticism from health and cancer experts in recent years, but until semidelivery became possible we had no direct way to learn what effect an expectant mother's smoking might have on her child. In chapter three we mentioned a study by Dr. Karlis Adamsons which might apply when we face the reality of substitute mothers. That study has even more immediate application in terms of regular pregnancies. In Dr. Adamsons' study forty monkeys were semidelivered during their gestation periods. After the fetus had been partially removed from the womb, catheters were attached to its chest, jugular vein, and stomach. The fetus was then returned to the womb and monitors attached to the tubes coming from the fetus through the mother's abdomen. Precise doses of nicotine were injected into the mother's bloodstream, in this case one milligram per 2.25 pounds of body weight is approximately the equivalent of a mother smoking a package of cigarettes, depending on how much the mother inhales. Injected nicotine is no different from that entering the body as smoke, and within minutes both reach the fetus. The monitors measured the effects on the monkey fetus with great precision, showing that the nicotine reduced the blood pressure and heart rate. The monitors also revealed that nicotine produces acidosis, a high level of acid in the blood which hinders the passage of oxygen to the fetus. But most crucial, according to Dr. Adamsons, is the depressing effect on the sympathetic nervous system of the fetus, for this system controls such vital involuntary functions as breathing, heartbeat, and fright.

What happens when a thirty-one-year-old wife, with Rh negative blood and in her seventh month of pregnancy, tells her physician, "I want this baby. I don't care if it's deformed; even if you have to cut off a leg I still want it."? After eight miscarriages, the psychological and emotional impact of yet another failure could easily prove disastrous. Despite repeated attempts to postpone a premature labor with the drug Halothane, the ninth miscarriage was less than a breath away. The attending fetologist, Dr. Stanley Asensio of the University of Puerto Rico School of Medicine, then faced a decision which hopefully will become unnecessary in the years to come for Rh negative mothers (because of Rho-GAM), even though it will likely become commonplace in other contexts. Dr. Asensio's first venture with human semidelivery lasted three hours, during which the woman's child was given a complete blood transfusion. In earlier attempts with human semidelivery at Yale and Columbia-Presbyterian Medical Center in New York City the operation was successful but resulted in premature labor, often within hours of the operation. The womb is very reactive and once tampered with it easily enters into labor contractions. The drug Halothane will, in some cases as in this one, forestall the premature delivery. Three weeks after the historic operation, the patient began to pass amniotic fluid through the birth canal, a sign either of the onset of labor or a break in the fetal membranes with its risk of infection. After an agonizing examination of all the circumstances, Dr. Asensio decided on a Caesarean section. Anemic and one month premature, the baby girl soon overcame her handicaps and today, four years later, she is a normal and healthy child in every respect.

In another later attempt to thwart the debilitating effects of invading maternal Rh antibodies, Dr. Asensio operated on a twenty-week-old fetus that weighed little more than a pound. Again the fetus was given a complete blood transfusion, this time through a catheter inserted in its jugular vein during its half-hour preview of this world. By the fall of 1969 Dr. Asensio had ventured into the field of semidelivery and prenatal surgery five times, with three successes. In one of these successes, the mother went into premature labor with her thirty-week-old fetus a few hours after the operation. This time Halothane was bypassed and another experimental technique tried. Needing only a week or two more before they could safely induce delivery, Dr. Asensio ordered daily injections of 20 percent alcohol for the mother. "Completely drunk, singing, happy and free of pain," the woman delivered a normal child eleven days later when her booster shot was withheld. Pioneered by Dr. Fritz Fuchs of Cornell University Medical Center, the alcohol treatment is still in its experimental phase, though in this case as in others it proved to be an effective retardant for premature delivery.

Implicit in prenatal monitoring is the desire to remedy abnormal situations and perhaps even improve on normal conditions in the development of the fetus. Few people raise moral or emotional objections to the former purpose. When amniocentesis reveals a dangerous surplus or shortage of a particular enzyme, few parents would object to the doctor administering remedial drugs. If amniocentesis revealed a mongoloid condition, few if any parents would object to the physician trying to prevent the presently inevitable mental retardation. Intrauterine photography and other monitors may reveal certain sex-linked defects, hemophilia, delayed or ambiguous sexual development, certain types of muscular dystrophy, and the congenital blindness and mental deficiencies associated with pseudoglioma among other deformities. Dr. George E. Burch has predicted that by 1971 or 1972 it will be possible for physicians to administer antibiotics to a fetus in the womb to halt infections before they damage the fetal heart. Again who could object to these attempts to control or remedy such deforming conditions?

The sensitive issue comes when you ask whether the doctor should try to improve on a normal fetus by administering certain drugs, or whether an abnormal fetus which cannot be helped should be aborted. Manipulation during semidelivery poses even more serious questions. "If you go to the early fetus, it is pretty fantastic what you can do—like take an arm off one animal and put it on another, change sex or even transplant heads," is the warning of Dr. Ronald E. Myers, chief of the National Institutes of Health Laboratory of Perinatal Physiology in San Juan. But will society tolerate or permit this venture? In the Harris *Life* survey, 58 percent of those surveyed were in favor of using genetic manipulation and other techniques to avoid having a retarded baby. But when asked whether similar techniques should be used to improve the human race, only 21 percent favored this while 57 percent opposed such experiments.

Dr. Frederick Scheidman has carried out an experiment in his laboratory at the University of Minnesota which gives us some idea of potential developments in this area of fetal manipulation. He has used a common tranquilizer, reserpine, for injection into baby chicks while they are still nestled inside their shells. Sometime before hatching, the fetal chick's brain is imprinted with certain instinctive behavioral patterns. One of these shows up right after hatching when the little chick immediately follows the first moving object it sees. In nature this is his own mother, and the imprinted behavior assures that the new chick will be around to learn from mother hen all it needs for survival. This prenatal imprinting is so strong that even a moving mechanical object will become an irresistible "mother" for the newly hatched. Somehow reserpine prevents this prenatal imprinting so that a chick treated with the drug before hatching will wander around aimlessly even though mother hen is within easy view. The drug also impairs the ability of the newly hatched chick to learn. Placed in a simple maze with a glass partition separating it from food on the other side, a normal chick will quickly find the side passage leading to the food. A drug-treated chick will peck away at the glass until it dies of starvation without ever trying to go around the side.

Prenatal mind modifiers, brain improvements, and similar manipulations of the fetus will be distinct possibilities for man in the next generation or two. In some limited degree they are already a reality. Particularly in the area of human sexuality, their present reality and future potential are, depending on your view, demonic or godlike. We have already spoken to some extent about selecting the sex of our children, but the problem is much deeper than that superficial aspect. Yet the question of fetal human sexuality offers a deep insight into the whole question of man and his powers to manipulate and even direct his future nature. Understanding the developmental *process* involved in the emergence of human sexuality during fetal life is crucial, because few people outside of some biologists, doctors, and psychologists are aware of the *six basic aspects of that development and their interrelationship.* Furthermore, understanding this prenatal emergence of human sexuality is critical for any understanding and appreciation of adult sexuality.

Sexuality has always been a mystery and puzzle for man. Married couples want an heir and often the firstborn male child takes on a crucial importance in a marriage. Animal breeders often prefer one sex over the other. Over the centuries both groups have tried a variety of techniques to insure the proper and desired sex of offspring. In the days of ancient Greece, Empedocles argued that male children are begotten in warmer wombs while female offspring are conceived in cool wombs. Anaxagoras disputed this, claiming that males were conceived on the right side of the womb and girls on the left. Aristotle preferred the temperature regulation of sex because obviously warmth means strength and masculinity. Young hot-blooded, vigorous males, human and otherwise, beget more male offspring. Older men tend to produce mostly girls because they are tired and worn out. However, if an old man has intercourse on a hot summer day his prospects for producing a boy are greatly increased, as are the chances for a girl when a young hot-blooded male mates on a cold winter night.

The quasi-scientific speculations of the Greeks, humorous as they appear, contrast very favorably with the superstition and myth of later centuries. Amulets, spells, mysterious magical concoctions, and astrological consultations dominated the medieval concern over prenatal sex. Compared with this later appreciation, even the primitive Swahili of East Africa come up as rather practical manipulators of prenatal sex. If a Swahili woman works hard during pregnancy, a boy will be born to her; if she is lazy, the offspring is a girl. Furthermore, if she appears to be working hard during pregnancy and still bears a girl, the father can always accuse his wife of secretly cheating by slipping off for a nap when everyone thought she was working hard.

We have seen several not too successful attempts to predetermine or select the sex of an animal or human by sorting out the sperm or aborting those of the undesired sex four or five days after conception. However, once the genetic sex is set by fertilization and a pregnancy is begun, two reasons might prompt a parent to want to know the sex of the expected offspring. Curiosity could raise the question, or a concern for an expected abnormality, because some sex-linked defect found in the family history raises the question of a possible therapeutic abortion. If the parents know of hemophilia in their family, a normal offspring might well depend on the sex issue. A girl might be a carrier for hemophilia but perfectly normal, while a male offspring could run the serious risk of being a hemophiliac.

Until recently the only way to learn the sex of a fetus early in pregnancy involved the risk of amniocentesis. Cells sloughed off by the fetus can be withdrawn in a sample of the amniotic fluid and examined for the Barr-body, or inactivated X chromosome, characteristic of females. In 1969 a new and much safer technique was developed by physicians at the University of California Medical Center in San Francisco. The research team, Dr. Melvin Grumback, Dr. Felix Conte, and Dr. Janina Walknowska, originally set out to find the causes behind the mother's failure to reject the fetus as a foreign body. In careful analyses of blood taken from pregnant women, the team occasionally found white blood cells, lymphocytes, containing XY chromosomes instead of the expected XX. Selecting twenty-one such women for more detailed study, they found that nineteen of these later gave birth to boys. Cell samples from these women had contained at least two XY lymphocytes, while the two false predictions were made on the basis of a single XY lymphocyte, in the cells examined. These scientists then suggested that the placenta does not, as we once believed, screen out all white blood cells from the fetus, preventing their passage into the maternal bloodstream. Some fetal white blood cells apparently do cross the placenta with histocompatible antigens that allow the mother to accept the fetus without rejection.

"We started out the study thinking a cell count of 400 would be enough to find an XY cell, if it were present," Dr. Conte commented in a report for *Science News*. "But we soon upped this number to 700 and then to 1,200." The team now feels that examining 1,200 cells from a pregnant woman's blood will allow them a 95 percent chance of finding at least two lymphocytes with the XY chromosomes if the child is a boy. When two such male cells are found, chances are the child will be a male; if they are not found, the indirect indication is a girl. So far, predictions have proven accurate as early as fourteen weeks after conception. This is a very tedious, time-consuming procedure, but if present attempts to program computers to scan cells and perform karyotypes are successful, the technique may become routine in many hospitals. A computerized study of the chromosomes might also include a search for chromosome abnormalities, as for instance an extra twenty-first chromosome which results in mongoloid idiocy.

If this were uncovered, the physicians might advise amniocentesis to obtain more fetal cells and a more certain indication of possible abnormalities.

"Male and female he made them,"—or so they say!

And would to heaven it were that simple for the geneticist, just a plain XX chromosome complement in every cell of a girl's body and an XY for a boy! Yet one in every two hundred Caucasians suffers from a genetic sexual abnormality, at least according to the few scientific studies carried out so far. It was only in 1956 that we learned the exact number of chromosomes, forty-six, in the human cell from the work of Drs. J. H. Tijio and Albert Levan at the Institute of Genetics in Lund, Sweden. Prior to that time humans were thought to have forty-eight chromosomes. Once the correct count was established, chromosome studies on human material rapidly accelerated. Among the forty-six chromosomes, arranged in twenty-three pairs of which one is either XX or XY, several types of abnormalities can pop up. A person may lack a single chromosome, monosomy, and end up with only one X chromosome and no Y or second X. In 1938 Dr. Henry Turner described the results of this condition; it produces abnormally short, sexually underdeveloped women with possible skeletal and heart deformities. Another variation in nuclear or genetic sex is known as trisomy. This involves an extra sex chromosome besides the normal pair, thus, XXY, XXX, or the very controversial XYY. A person might also have a complete double set of paired chromosomes, including two extra sex chromosomes so that his cells contain forty-four paired body chromosomes plus four X or two X and two Y chromosomes. This is known as tetrasomy. Besides these changes in number, deletion can occur in the sex chromosomes so that a piece containing several genes is lost or translocated to another chromosome. Finally, you can end up with a mosaic individual. This occasionally turns up in butterflies and birds where part of the body is definitely female in genetic makeup and appearance while another part is typically male. Robins have been found, for instance, which have male plumage on one part of their body and female coloring on another part. Women have been known to have some cells with only one X chromosome, others with the normal XX, and still others with XXX. Depending where the abnormal cells show up in the body and how many of them there are, the symptoms for this mosaic abnormality vary greatly.*

Newspaper headlines in recent years have aroused public interest in one very practical and controversial application of nuclear sex to the courtroom. "Did chromosome spark murder?" "'Chromosome Slaying Trial' Begins in Queens." "Richard Speck Pleads 'Genetic Insanity.'"

Geneticists have been studying the effects of an additional Y chromosome in certain males since this phenomenon was first reported in 1961. By 1965 several psychologists and sociologists suspected a possible relationship between a male with an XYY chromosome pattern and aggressive violent behavior. The indication was that individuals with an additional Y chromosome are unusually tall men, mentally slow, troubled with acne and pimples, and with strong antisocial and aggressive tendencies. In a 1965 report on "Aggressive Behavior, Mental Subnormality and the XYY Male," in *Nature* Dr. P. A. Jacobs studied 197 mentally subnormal males with dangerous, violent or criminal tendencies in a penal institution. Twelve males with chromosome abnormalities were found and seven of these had the XYY pattern: one had XXYY. Thus eight out of 197 had an extra Y chromosome. In 266 randomly selected newborn males and 209 adult males he found no case of an XYY individual. An additional 1,500 males examined for other reasons revealed only one XYY individual. However, when some British psychologists tested seven XYY patients they did not appear significantly different from normal men in either intelligence or tested hostility. Other scientists have found that testosterone, the male hormone, is often at a higher level in XYY males.

---

* The nuclear sex of women with Turner's syndrome may have a variety of combinations which the geneticist records as follows: XO, X″X″, X″x″, XO/XX, XO/XXX, XO/XX/XXX, XX/XXX, and XO/X″x″. In the masculine parallel to Turner's syndrome, known as Klinefelter's syndrome, the genetic combinations for the sex chromosomes in each cell include: XXY, XXXY, XXXXY, XXYY, XY/XXY, XXY/XX, XXXY/XXXXY, XY/XXY/XXYY, and XY/XO/XXY. These chromosome variations are also marked by different conditions in the Barr-body and the drumstick formation in white blood cells, leucocytes, both of which are usually indications of female sex. The / in the above notations indicates a mosaic with the different types of nuclear complements separated by the slash.

With perhaps 200,000 American males walking around today with an XYY chromosome complement the issue becomes even more controversial. Murder trials in France and Australia have used "genetic insanity" as a defense plea. In the Paris trial the accused received a light and reduced sentence; in Australia, the verdict was acquittal because of "genetic insanity." In a Queens, New York, murder trial the defense attorney pleaded not guilty for his client by reason of an insanity resulting from chromosome imbalance. The additional Y chromosome, he maintained, rendered this client the guiltless victim of an overdose of male hormones which accounted for his aggressive and brutal rape-murder of a divorcée in August of 1968. Under oath a geneticist would not agree that the extra Y chromosome was necessarily related to antisocial or criminal behavior. "It can contribute, but by itself, it would not be the factor," was the testimony of Dr. Edward Shutta, geneticist at the Downstate Medical Center of the State University of New York.

Thus far, the evidence is far too incomplete for any definitive verdict to be reached. As Dr. Gerald Clark told a gathering of the American Psychiatric Association in May of 1969, the frequency with which XYY men are involved in violent crimes and other antisocial behavior "may not be appreciably different from the average citizen." Commenting on the issue with regard to the legal controversy and the state of genetics and psychology, Dr. James Hamilton, also of the Downstate Medical Center, concluded that gathering and understanding all the evidence will be a "long and difficult process." As an instance of the wide range this evidence must embrace we could cite Dr. Hamilton's own tests of normal male killifish in competition for mates with supermales (YY). When XY and YY males were set in one to one challenge for a single female, 137 contests were won by the supermales and only eighteen by the normal males. Whether this superiority in aggressive behavior is due directly to the extra Y chromosome or indirectly to the additional male hormones produced is at this point unknown.

If nuclear or genetic sexuality is sometimes confusing, hormonal sexuality is even more so.

In the 1920's and 1930's little if anything was known about genetic sexual ambiguities outside of the XX and XY patterns. At the same time the science of internal secretions, endocrinology, was in an even more primitive state on the question of human sexuality. It was thought, for instance, that the male and female sexes were determined hormonally by a single male and a single female hormone produced respectively by the testes and the ovaries. This theory was as simple and clear as the unadulterated view of XX and XY was for the geneticist at the time. But with the early work on the steroid hormones the situation soon became confused and complicated. In 1928 Brouka and Simmonet found that testes extracts produced specifically female effects on the vagina. Six years later Zondek found the testes of a stallion to be the richest source of female hormones of all tissues examined at that time, whether from male or female animals. The urine of a stallion was found to contain much more estrogen, female hormone, than the urine of a nonpregnant mare. A clue to the ambisexuality of the human came with the discovery of both male and female hormones in the urine of women and men alike, the only difference being in the proportions. Adding to the confusion was another unexpected finding, that androgen and estrogen come not only from both the testes and ovaries, but also from the cortex of the adrenal glands which were supposed to have nothing at all to do with human sexuality. As Professor A. S. Parkes summed up the situation in his collection of essays, *Sex, Science and Society:*

> The present wonder, therefore, is not that intersexual conditions occur, but that the balance of endocrine factors usually comes down on one side or the other to produce a recognizable male or female—perhaps in these days I should say a more or less recognizable male or female.

Human sexuality is a PROCESS, a sequence of developmental events which begins with the genetic code of XX or XY but involves a *lifelong becoming*. Exactly what it is that activates the differentiation of the primordial reproductive glands, the gonads, to go down one or the other channel is unknown today. In some way the XX genetic code triggers the cortex or surface of the primitive undifferentiated glands to form ovarian tissue while the XY code activates the inner core of the same glands to form seminiferous tubules and testes. Once the gonads have specialized, they themselves become a prime factor in regulating the differentiation of the remaining elements of the reproductive system and human sexuality. A professor of medical psychology at the Johns Hopkins University, Dr. John Money, has challenged many time-honored beliefs about sexual identity. Some of these are implicit in the multiphased process view outlined in my adaptation of his systematic analysis as follows:

1. *Genetic or nuclear sexuality* as revealed by indicators like the sex-chromatin or Barr-body, a full chromosome count and the leucocytic drumstick;
2. *Hormonal sexuality* which results from a balance that is predominantly androgenic or estrogenic;
3. *Gonadal sexuality* which may be clearly ovarian or testicular, but occasionally also mixed;
4. *Internal sexuality* as disclosed in the structure of the internal reproductive system;
5. *External genital sexuality* as revealed in the external anatomy, and finally;
6. *Psychosexual development* which through the external forces of rearing and social conditioning along with the individual's response to these factors directs the development of a personality which is by nature sexual.

Several instances have recently made headlines in the popular press where apparently a contradiction occurred between a person's genetic and gonadal sexuality, or between his gonadal/internal/external sexuality and his psychosexual development. In simple terms, the process of sexual development started off in one track and for some unknown reason switched horses in midstream.

A classic example of this is Miss Ewa Klobukowska, a Polish entrant in the 1967 European Cup Women's Field and Track Competition. Co-holder of the world's record for the hundred-meter dash at the time, she was ruled ineligible because a chromosome count from cheek cells revealed "one chromosome too many." The twenty-one-year-old miss, it seems, was a genetic mosaic with some cells in her body containing a single X chromosome and the others containing a single X and two Y chromosomes. She apparently started off with a set of genes that should have produced a male: XY. Early in her development an unequal distribution of chromosomes at cell division sent three sex chromosomes to one daughter cell, XYY, and only one to the other daughter cell, X. A normal cell would have duplicated its XY complement, with the XXYY set then dividing equally, one XY pair going to one cell and the other XY to the other cell. In Miss Klobukowska's case the unequal distribution resulted in a person whose physical appearance was that of a sexually immature woman, quite tall and with underdeveloped breasts. If such a condition can be diagnosed early enough during pregnancy, it might soon be possible for doctors to correct the confused or contradictory development and set the process back on its proper genetic path.

Writing in the *Journal of the American Medical Association* Dr. Keith Moore of the University of Manitoba has argued, and rightly so, that "no single index or criterion can signify the appropriate sex for an individual." Dr. Moore questioned rather severely the "unreasonable dependence" on chromosome patterns from smears of cheek cells in determining athletic eligibility. In his editorial he pointed out that such tests "are merely indicators of sex, and should not be used as absolute criteria of sexual identity."

In mammals and man ambisexuality, the appearance of characteristics of both sexes in each animal of the species, is evident also in the early development of the internal and external reproductive system. One might expect that reproductive systems as different as those we find in the adult man and woman would be sharply different in their origin and earliest appearance. Yet every mammal, including the human, starts off from a sexually indifferent stage where *two* parallel primordial reproductive systems make sexual determination on the basis of anatomy impossible. Internally and externally the male and female fetuses are indistinguishable. A pair of primordial gonads or reproductive glands are present as are two parallel pairs of tubes, the mesonephric (middle kidney) ducts with their many small tubules reaching out to the gonads and the unbranched Mullerian ducts which parallel them. Both pairs of ducts lead from the region of the primordial gonads to the cloacal opening to the outside. The external anatomy of this indifferent stage is limited to a genital tubercle and the swellings and folds of the cloacal opening, which surround the urogenital opening.

In a male, once the chromosome complement has triggered the inner core of the primordial gonads to proliferate into testes and produce a predominantly male balance of hormones in the blood, the indifferent system also starts down the channel toward maleness in other respects. The mesonephric ducts connect with the new testes and the Mullerian ducts degenerate except for certain vestigial remnants.

Externally, the genital tubercle becomes the penis. The genital folds behind it form the lower part of the penis while the genital swelling forms the scrotum which will encase the testes when they descend.

In the female, once the chromosome complement has triggered the outer skin of the primordial gonads to proliferate into ovaries and produce a predominantly female balance of hormones in the blood, the indifferent system also starts down the channel toward femaleness in other respects. The mesonephric duct and tubules degenerate except for a vestigial remnant and the Mullerian ducts become the fallopian or oviducts. Near the cloacal opening the Mullerian ducts fuse to form the uterus and vagina.

Externally, the genital tubercle becomes the clitoris. The genital folds and swelling become the minor and major lips of the vulva respectively.

A few examples from different animal experiments may help clarify the deep interrelationship and interaction of hormonal and anatomical sexuality in terms of the basic ambisexuality of mammals and man.

In 1934 Wiesner proposed a theory to explain the relationship of sexual hormones and anatomical development in the mammals which held that female structures basically develop independently of the sexual hormones, while the fundamentals of the male system depend on the presence of male hormones. Support for this theory has come from a variety of experiments, so that it is now generally accepted. If, for instance, male genital primordia from a rat or mouse fetus are cultured *in vitro* with either the male hormone or embryonic testes, the mesonephric ducts continue to grow and differentiate while the Mullerian ducts degenerate. When the hormones or testes are left out of the culture medium, it develops typically female structures. When female genital primordia are cultured outside the rat or mouse, the female duct system continues to develop whether or not the female's hormones or ovaries are present. Furthermore, castrated and normal female and castrated male rats and mice all develop essentially the same.

Freemartins offer another interesting insight into the relationship of hormonal and anatomical sexuality. A freemartin is a female calf which has as a twin a male calf. Although her male twin is physically normal, the freemartin calf has an undeveloped or malformed reproductive system. Years ago Professor F. R. Lillie, a giant in early embryology, showed that the placentas to such fraternal twins overlap in the womb so that the male hormones pass into the blood stream of the female calf and divert its normal sexual development.

Finally, sheer logic forces us to accept a theory such as Wiesner's if for no other reason than that without it we would all theoretically end up female, with no males anywhere. The female hormone does pass through the placenta from mother to fetus. If it triggered basic female structures, then every fetus would turn out anatomically female whatever might be its genetic sex, since the female hormones are always present. Logic then demands a system in which the male develops under the influence of his own male hormones and the female automatically without hormonal influence.

The secondary sexual characteristics, on the other hand, appear much later in life and are clearly under the influence of hormonal control. These secondary traits include changes in the voice, body hair, maturation of the reproductive system, and the female breasts.

Considering the ambisexuality of man during early fetal life, one can seriously ask whether it might not be possible to reverse the sexual development of an individual.

In birds the female has a rudimentary testis-like structure which may enlarge and produce male hormones if the functioning ovary is removed or destroyed by disease. This is why old hens occasionally grow spurs and develop other male characteristics and begin to crow like a rooster. One case has been reported by Professor Crew in which a hen, after producing normal eggs, gradually developed the behavior and appearance of a rooster, mated with some hens, and finally became the father of two chickens. This natural reversal of sex can be duplicated experimentally without much difficulty. It should be noted however, that what is being reversed is the anatomical and hormonal sex, not the genetic or nuclear sex which remains as it was at conception. Among some amphibians (frogs and toads), fish, and other lower vertebrates, spontaneous reversal of anatomical and functional sex is relatively common.

Experimentally, anatomical sex reversal has been reported in the rat by Steinach and others. Two research biologists working for the Schering Company in Berlin, F. Neumann and M. Kramer, have injected pregnant rats with cyproterone acetate during the last half of gestation and found that all the males were born with vaginas. Soon after birth, these males were castrated and ovaries implanted in them. The transplanted ovaries not only produced normal eggs and a regular cycle of female hormones, they also produced female behavior so close to the normal that not even normal untreated males could detect any difference between transformed and ordinary females. Very similar results were obtained when the drug was administered shortly after birth to males.

Drs. Charles Phoenix and John Ruskow, at the Oregon Regional Primate Research Center, have tried injecting the male hormone, testosterone, into pregnant monkeys to check its possible effects on sexual development. When injected in the right amount at just the right time in development, all the newborn, both the genetic males and the genetic females, were conditioned as males in all their social behavior, in play patterns, dominance within the group, mating and grooming. In 1959 Phoenix and some co-workers bred hermaphroditic guinea pigs, again by administering testosterone to pregnant mothers. Some of the daughters were born with male external features and female internal structures. Yet in behavior these experimental hermaphrodites were closer to normal males in their behavior than to normal females. The researchers attribute this to an organizing effect of the male hormone on the brain, particularly the hypothalamus, which controls psychosexual behavior.

Late in 1954 Dr. Alan Fisher began some provocative experiments in the laboratory of D. O. Hebb at McGill University with psychosexual conditioning. His hope was to produce direct chemical stimulation of specific brain cells. At the time scientists knew that, in the vertebrate brain, the hypothalamus, the floor and sides of the cerebral hemispheres above the pituitary gland were involved in such primary drives as courtship behavior, the care of the young, eating, and drinking. Since the hormones, especially the sex or steroid hormones, also affect these behavioral traits, Fisher decided to inject them into specific regions of the hypothalamus. Within seconds after an hypothalmic injection with testosterone, a male rat became very restless. Placed in a cage with a female rat not in heat, Fisher expected him to make advances to the young lady, though normally a male rat will not be attracted to a female unless she is in a receptive state. Prompted by the male hormone, however, the rat wrote his own script. He grabbed the female by the tail and dragged her across the cage to another corner. When she scurried away, he dragged her back, finally picking her up by the nape of the neck instead of grabbing her tail. The male rat, injected with the male hormones, was really confused; his behavior was typically maternal. To test this theory, Fisher placed some nesting material and a couple of newborn pups in the cage. Immediately the male rat began to build a nest and tuck the pups in like any solicitous mother would do. Half an hour later, the maternal behavior began to wear off. A second injection, and papa was once more mama. Repeated injections, though, brought the rat into real confusion. His behavior was still maternal, but disordered and haphazard. Granted an overnight rest, however, the male hormone would again produce normal maternal behavior the next day. When the male hormone was injected slightly to the side of this first midline injection, the effect was strikingly different. Then the behavior was typically male. When a point midway between the center and side injection locations was used, the effect was a "curious combination of maternal and male behavior. They took care of the young and at the same time tried to copulate with any partner available." In several cases a carefully injected male would even try to mate with a nonreceptive female while he was carrying a pup in his mouth! Testosterone, the male hormone, then can mimic the action of the female hormones, and vice versa!

"For years," Etienne Wolff has noted, "we have also been able to change the sex of an individual by injecting hormones into the embryo." There is no predicting, though, what might happen were this treatment to be tried on man. However, in attempts to remedy an apparent contradiction between anatomical and psychological sexuality in a single human, this work suggests some valuable insights into the transsexual phenomenon and its possible therapy. This sensationalized phenomenon first received public notice with the cases of Christine (Christopher) Jorgensen and Roberta (Robert) Cowell, an English airman, and more recently with the opening of a clinic to treat transsexuals at Johns Hopkins Medical School.

The brain tissue itself is known to be female in its fundamental psychosexual orientation. This applies particularly to the hypothalamus buried at the base of the brain above the pituitary and between the two cerebral hemispheres, about in the center of the head. Preservation of self and of the species is the prime function of this oldest portion of the mammalian brain which we share with all other mammals, even the lowest forms. It controls respiration, pulse rate, and blood pressure, governs the salt balance in the blood and regulates the temperature of the body, besides guiding certain built-in instinctual responses such as hunger, thirst, aggression, fear, play, sleep, waking, and sexual behavior. Fundamentally female-oriented, only the presence of an added factor, the male hormone, will allow this tissue a male psychosexual development. In some of the experiments mentioned above a genetic female was diverted into a psychological male by the administration of the male hormone, while the absence of the male hormone in a genetic male allows the basic female orientation to come through unmodified. Seymour Levine, another researcher in this field, has suggested that the basic female orientation of the brain and a conflicting anatomical development may be an important factor in homosexuality. For this reason some researchers have argued that it may be much easier to change human males into psychosexual females than vice versa. The treatment of transsexuals seems to bear this out, as all the cases thus far reported have been anatomical males with a contradictory female psychology that can only be resolved by surgically supplying the confused individual with a partially functional female anatomy. In pondering the possible parallel between the basic female orientation of the brain and anatomical development in the subhuman mammals on one hand and our human situation on the other, Dr. John Money makes an interesting comment in an essay on "Psychosexual Differentiation": "In view of the alleged higher incidence of psychosexual pathologies in males, it is conceivable that masculine psychosexual differentiation is more difficult to achieve than feminine, and is more vulnerable to error and failure."

All of this brings into focus yet another perplexity raised by the new embryology. In the preceding chapters we found that the new trends in reproduction have brought some fundamental and serious confusions into our traditional use of such terms and concepts as parent, father, mother, son, daughter, adultery, fornication, incest, and marital fidelity. Now we find that even the most fundamental conception, that of male and female, is far from clear when we try precise and limited definitions. Perhaps our only salvation is to realize the limitations of terminology and conceptualizations, and rely on the old proverbial "common sense" illumined certainly by a respect for the facts and ambiguities of our scientific knowledge.

Genetic mosaics, which we mentioned above, occur rather frequently among the birds and insects, particularly the butterflies and moths, but an example from the common fruit fly, *Drosophila*, should provide an amusing conclusion to this disturbing survey of fetal sexuality, its occasional ambiguities, and its possible manipulations. When a gynandromorph occurs in this insect the mosaic of male and female tissues can cause real complications in the behavioral patterns. In the *Drosophila* courtship dance the male holds his wings vertically and vibrates them rapidly, while the female spreads hers quietly to the side. In the case of gynandromorphs, one side of the body may be genetically, physically, and psychosexually male and the other side female. When this type of fly is placed with normal flies of both sexes, he-she gets very confused: the male side of the body reacts by quivering a vertical wing while the wing on the other side hovers motionlessly and horizontally. Similar sexual mosaics have been observed in *Hebrobracon,* a tiny parasitic wasp, in which the front part of the body including the brain is one sex, while the abdomen and reproductive structures are of the other sex. The wasp may then try to court and mate like the psychological male it is, even though its reproductive system is functionally female. Again the confusion makes for an impossible, if humorous, combination to watch.

Every prospective father and mother has one prime concern: that the child they have conceived and hope to bear will be normal and healthy. In the delivery and waiting rooms this concern may yield momentarily to fear for the mother's life, but the fear of giving birth to an abnormal child, with its many psychological and emotional faces, is more lasting. The birth of a defective child can undermine or even destroy the self-esteem of a new father or mother. It can bring strong tensions, shame, and recriminations into the husband-wife relationship. It can even touch our basic instincts for human survival and compassion to the point where we are tempted by ancient remedies. Primitive peoples and even Western civilized man of not so long ago regarded a defective child as a punishment of the gods, a divine reprisal for an ill-spent youth, an ill omen, or the product of some demonic curse that must be purged by death.

Of the approximately quarter-million children born each year in the United States with important birth defects, 20 percent of these are due to specific inherited genetic defects, while another 20 percent can be traced to environmental factors like poor nutrition, the lack of prenatal care, X rays, and German measles. The remaining 60 percent are due to interactions of genes and environment, and include the effects of chemical mutagens which alter the chromosomes of mother and child, or interfere with gene action. Such would be the effect of excessive use of sugar substitutes, LSD, and the tranquilizer thalidomide. According to Dr. Kurt Hirschhorn, president of the American Society of Human Genetics and professor of pediatrics and genetics at Manhattan's Mount Sinai School of Medicine, every human being carries from three to eight defective recessive genes in his hereditary information. Geneticists estimate that at least 5 percent of all Americans need or could profit from some kind of professional genetic counseling.

The science and profession of genetic counseling is relatively new in the field of medicine, though questions of eugenics and a variety of sane and insane proposals have flourished since the turn of the century. Only recently has genetic counseling become anywhere near a practical science. Over fifteen hundred hereditary defects are known to human geneticists today. Yet methods of detecting these defective genes while the child is still in the womb or of predicting the odds for the conception of a defective child by a careful study of the two prospective parents' family histories are effective only with perhaps three dozen abnormalities. The typical genetic counselor today is not even granted the status of recognition by his medical associates. Almost invariably he has come to his new profession by accident. In many cases he is a trained professional in pediatrics or gynecology who became interested in helping parents with a high frequency of abnormal children in the family. Or he may be a professional geneticist, fascinated in his later career by the almost virgin field of human genetics. He might even come into genetic counseling through the back door of some laboratory research project on a special enzyme deficiency or metabolic disturbance.

Until recently a lack of both specialized training and of money for research and facilities has greatly retarded the progress of a service science which may very well become a vital concern for parents by 1975 or sooner. Yet despite the obstacles and the lack of funds, the number of American hospitals providing services to prospective parents in genetic counseling has grown from 20 in 1955 to 114 in early 1969. Mount Sinai Hospital in New York, Baltimore's Johns Hopkins Hospital, George Washington University Hospital in Washington, D.C., and Children's Memorial Hospital in Chicago have major genetic counseling services. In 1964 the state of New York underwrote the cost of establishing a center for genetic counseling at the Albany Medical Center. The first major clinical center specifically designed with research and laboratory facilities to detect inherited fetal defects was recently set up at the Johns Hopkins University Hospital with support from the National Foundation. We have not even begun to face the major need for men and women trained in human genetics and in counseling techniques. Sarah Lawrence College is the only educational institution even to enter into planning a program for training young people in this new field of medical technology.

Another major factor holding back the expansion of genetic counseling as a regular practice is the cost factor for the patient. There is usually no charge for the advice given by counselors, as they are generally members of a clinic staff and on salary, but the laboratory fees can vary greatly. If the laboratory tests fall within the scope and interests of a staff member who happens to be working on a special project financed by some grant, there is often no charge at all. But this is frequently not the case, and a basic chromosome analysis usually runs in the neighborhood of $125. A few medical centers charge ten dollars or less, but only one state health department offers this service free, New York through the State Health Department's Birth Defects Institute in Albany.

In the past doctors have been limited to examining the family histories of both parents in an effort to locate instances of relatives with definite inheritable defects. They might then in some few instances such as hemophilia or mongoloidism inform the parents that they had one chance in four of having a defective child, or that the odds were fifty-fifty. But the statistics were often difficult to uncover in the first place, and then to present in a simple but valid way to parents who find the mathematics often incomprehensible or too abstract. With some of the prenatal monitoring techniques, a whole new world has opened up. "Intrauterine detection [of defective fetuses] brings a new dimension to genetic counseling. The physician may now inform the parents that they will have either an affected or a normal child," according to Dr. H. L. Nadler at Chicago's Children's Hospital. Half a century ago infectious diseases caused about 60 percent of childhood deaths and only 2 or 3 percent could be blamed on genetic disorders. Today the situation is completely reversed with infectious diseases accounting for about 2 percent of childhood deaths while genetic disorders are now responsible for some 12 percent. This reversal is making genetic counseling even more important and urgent than it ever would have been in past generations.

Amniocentesis is a major tool for the genetic counselor. With it, he can tell whether a three-month-old fetus, scarcely an inch long and weighing about three ounces, has cystic fibrosis, galactosemia (the inability to metabolize a substance in milk which results in mental retardation), or will be a mongoloid idiot. When such diseases ran in a person's family, parents only recently had to wait an agonizing nine months to learn whether their child would escape unharmed to enter this world as a normal healthy baby. Amniocentesis can now answer the question in three months. In some cases the news is reassuring, but in others it brings the added moral decision of whether or not to ask for a therapeutic abortion or to face the emotional and financial problems of raising a retarded child. As of 1969 abortion to prevent the birth of a defective child was legal in only seven states, Arkansas, Colorado, Georgia, Kansas, Maryland, New Mexico, and North Carolina. Many of the states and countries now considering changes in their abortion laws have not considered at all the possible implications of genetic counseling on the new legal provisions. As usual we seem to be muddling along with our legal apparatus centuries behind our medical capabilities. As a result, few doctors now involved in genetic counseling and the occasional decisions this raises in recommending a therapeutic abortion as an "unjustified psychiatric trauma" for the mother are willing to discuss their cases or acknowledge their recommendations. At the present time our choices in genetic counseling are very limited, in most cases the choice is between therapeutic abortion and permitting the birth of a defective child. While amniocentesis will eliminate indecision on the abnormality, other types of prenatal monitoring, such as the intrauterine photographs of Lennart Nilsson and the Lileys, are going to make the decision to abort much more personal and individual, more agonizing and emotional. In the past it was very easy to regard the early fetus as something of an inert vegetable rather than as a vital, living human or pro-human individual.

Within a generation we may very well be able to inject viralcopies of specific genes to replace defective or missing genes in the fetus. Nobelist Arthur Kornberg has devised a way to duplicate a viral chain of four or five genes artificially in the test tube. This technique combined with the 1969 success of Harvard's Jonathan Beckwith in isolating for the first time a single specific gene, the lactase complex, makes it more than a mere probability that we will in the not too distant future be isolating specific viral genes for such proteins and enzymes as normal blood hemoglobin and insulin production, duplicate these artificially by the Kornberg method, and inject these artificial genes into the defective child where they can be incorporated into his cells and start producing the needed enzymes or proteins.

With each passing month our potential for making therapeutic abortion *a technique of man's primitive past* comes closer to reality. Almost every month at least one inherited disease is traced to the lack of a specific enzyme, which may soon be supplied by pills or injections to the deficient child even while he is in the womb. While abortion might be considered as a solution today, a fetus known by amniocentesis to have a sex-linked neurological disease which results in a type of cerebral palsy may soon be supplied with the deficient culprit enzyme, PRT. Similarly southern Eskimo and Athabaskan Indian babies born in Alaska, and Navajo Indian babies born in the Southwest with a rare disease that produces bluish lips and fingernails and a slate-gray skin, might be given by injection or pill the enzyme NADH-methemoglobin reductase which their bodies cannot manufacture. Another genetic disorder which may be diagnosed with amniocentesis and remedied by the administration of a missing enzyme is the rare Tay-Sachs babies, who show a fatal brain deterioration by their fourth year because they lack the key enzyme hexosaminidasea.

Ironically, genetic counseling can be a dangerous development in unexpected ways. A cleft palate and lip is not really a major crippling defect and plastic surgery can often remedy it. Yet when one prospective mother was informed that her chances of having a child with a cleft palate and lip were one in twenty, the woman resorted to an illegal abortion. The odds were very much in favor of a normal baby, about 95 percent so, yet somehow this optimistic diagnosis did not reach the woman and she panicked.

About one American Negro in every ten is genetically subject to sickle-cell anemia, a blood condition in which some of the normally doughnut-shaped red cells change their shape to that of a crescent or sickle under certain stress conditions. This change in shape reduces the cells' ability to carry oxygen and often leads to fatal blood clots in the brain and other vital organs. Centuries ago the genetic mutation responsible for this sickling trait evolved and spread through the negroid populations of malaria-ridden Africa. Similar anemias have evolved and spread among the Italians, Turks, Greeks, Sicilians, and Asiatic Indians in areas where malaria is common. The gene responsible for the rigid and quite fragile sickled cells would ordinarily have been eliminated or kept at a very low level in the population by natural selection, since children with the disease would die during their first year of life. But while the sickle-cell anemia is harmful, the malarial parasites do not infect a person with one normal gene and one recessive sickle gene. The sickle gene produces a type of hybrid vigor or heterosis in a malaria-ridden population. People with two normal genes die from the malaria, while people with two recessive sickle genes die of the anemia. But the hybrid thrives, at least where malaria thrives. In the United States malaria is rare and interracial marriages have reduced the percentage of the sickle gene in the Negro population. Nevertheless it is not uncommon for two Negroes, both of whom carry one sickle gene, to marry and have children. One-quarter of their children will not survive their first year while one-half of them will, like their parents, face the risk of death should they do some high mountain climbing, travel in a jet with defective oxygen supply, or undergo anesthesia during an operation.

In 1969 a simple prenatal test was developed by scientists at Ortho Diagnostics in New Jersey. A single drop of fetal blood, placed in a tube containing a chemical fluid, will turn cloudy within five minutes if sickle cells are present. The test costs less than a dollar and is more than 95 percent accurate. The accuracy, speed, and simplicity of this test make it an ideal screening procedure for large numbers of people. It brings into the realm of practicality a proposal many people considered incredible and totally absurd only a few years ago. In 1967 Dr. Charles F. Whitten on the medical faculty of Detroit Children's Hospital, suggested that we could control and even abolish sickle-cell anemia among the American Negro if individuals were informed that they are carriers of the disease and if they could be discouraged from marrying another carrier. Even if two carriers married, they could still avoid passing the defective gene to their children by using artificial insemination with donor sperm or egg, or by resorting to amniocentesis and therapeutic abortion for the defective children they conceive. Dr. Whitten has suggested that a program of genetic counseling be introduced in the high schools which would test every child and let each student know his genetic background. Coupled with the testing would be an educational program to explain the risks and possibilities in marriage to the students. Civil laws already require blood tests before marriage, chest X rays for teachers and food handlers. Is it beyond the conceivable to say that a genetic counseling program such as that proposed by Dr. Whitten may indeed become a reality in our school system within a few years? The case cited here contains the complicating factor of a disease limited to one segment of the population, where racial and minority emotions can be explosive, but there are other diseases for which genetic counseling programs are foreseeable in the near future.

Cystic fibrosis, with its mucous congestion of lungs and other organs, afflicts one in every sixteen hundred newborn babies. Since it is a recessive trait, both parents must be carriers for the disease to show up in a child. In this situation one in four of the children will have the disease and half of all the children will be carriers. If, for the next generation, we required therapeutic abortion for all fetuses discovered by amniocentesis to be carriers, we could "eradicate cystic fibrosis from the American scene," according to Dr. Hirschhorn. The medical profession obviously is not enthralled with this possiblity. An achondroplastic dwarf, another genetic defect, is the victim of a dominant gene and one-half of his children will be dwarfs. Thus artificial insemination with a donor's sperm might eliminate this defect entirely from the gene pool.

One of the most insidious inherited diseases known to science today is Huntington's chorea, insidious because its effects do not usually show up until around the age of thirty-five when most people have already established their families and passed on to the next generation whatever defective genes they might have. The disease leads to a gradual mental deterioration, uncontrollable jerking and flailing of the body and arms, and an irritable premature senility. In one case cited by a leading genetic counselor, a husband's grandmother, mother, and a couple of uncles died of Huntington's chorea, giving him and any child he might have a fifty-fifty chance of suffering the same debilitating premature senility and death. When informed of the odds, he refused to have any more children; the wife refused to use any contraceptive and the result was infrequent lovemaking, tensions, and a marriage in jeopardy. Adoption or artificial insemination might have helped, but as long as no method of detecting the disease in a fetus and relieving it is known such cases will prove serious problems for marriage counselors.

The mental retardation, short stature, flattened nose, and broad hands and feet which characterize the victim of mongoloid idiocy or Down's syndrome are generally due to an extra chromosome in the twenty-first pair. One out of every six hundred babies born in the United States is afflicted with this crippling disease. In some cases it is hereditary within families, being passed on by carriers who have a recessive gene which permits the extra chromosome to arise by unequal distribution or "nondisjunction" during cell division. In most cases it seems not to be an inherited trait but rather to depend on physiological conditions in the mother, which allow an abnormal formation of chromosomes in the egg. The odds of a married couple having a mongoloid child vary with their age. For a woman between fifteen and nineteen, the odds are only one in 1,850, for a woman in her early thirties, one in about eight hundred, and for a woman in her late forties, one chance in fifty of her having a mongoloid infant. Where the disease is inherited the statistical odds can be calculated and where it seems to depend on other factors, amniocentesis can often be used as a diagnostic technique.

The question of mongoloid children is a perplexing and pathetic one for the scientists as for the parents of such youngsters. Dr. James German, director of the Human Genetics Laboratory at the New York Blood Center and a professor at Cornell University Medical School, believes a clue to the cause behind older women being more prone to having mongoloid children lies in the waning sexual activity of middle-aged married couples. He has suggested that if marital intercourse is infrequent the egg or sperm may deteriorate somewhat before fertilization occurs. If a woman has intercourse every two days chances are that fresh sperm will be present to fertilize any egg ovulated within twenty-four hours. When intercourse is less frequent the sperm or egg may still retain its fertilization capacity even though its chromosome complement has started to deteriorate. Dr. German quoted the Kinsey Report in proposing his theory in *Nature,* showing that at twenty-five the average American married couple engages in intercourse 2.5 times a week and at the age of forty-five only 1.3 times weekly. In 1969 Professors Matsunaga and Maruyama of the National Institute of Genetics in Mishima, Japan, countered this claim by showing that there is no correlation between the Kinsey figures and the frequency of mongoloids, since before the age of thirty mongoloids occur at a fairly stable rate while after thirty their frequency rises exponentially, a curve which does not correspond with the curve for frequency of intercourse.

The rhythm method of limiting conception has often been criticized for its ineffectiveness, but recent studies by two eminent gynecologists, one Irish and one Italian, have suggested that it also carries with its practice a serious risk of fetal deformation. In a study of *spina bifida,* a failure of the spinal cord to close, Dublin's Dr. Raymond G. Cross found the frequency varies with religious persuasion; occurring in 2.8 births per thousand among Catholics, 2.0 per thousand among Protestants, and only 0.7 per thousand among Jews. Since few Protestants and very few Jews use the rhythm methods, Cross claims that the frequency of the deformity can be due to stale sperm or deteriorating eggs being fertilized and thus the variation in its frequency. Dr. Carlo Sirtori, of Milan, agrees with Cross's theory and argues that mongoloid idiocy and other birth deformities may also be due to the use of rhythm and the deteriorating condition of germ cells when fertilization occurs. Though no concrete evidence has substantiated this theory, both Dr. Emil Witshi, senior investigator in reproduction and embryology for the Population Council, and Dr. Alan Guttmacher, president of the Planned Parenthood League, feel the theory has some substance.

All of which brings us back to the basic question of whether we should and how we might best improve the lot of the fetus.

Dr. Hermann J. Muller's bluntly eugenic program for limiting reproduction to artificial insemination with selected frozen sperm is not likely to receive anything like general public acceptance for some years, although an increased voluntary use of artificial insemination, ova transplants, and frozen germ cells beyond present use in solving sterility is probable in the next decade. Defective genes could be reduced by legal restrictions of the right to reproduce, but this seems most improbable for some years to come since democratic Western man still treasures his "inalienable" right to reproduce at will far too much to permit its limitation voluntarily. Asexual reproduction using the normal cells from one spouse might offer a solution in the future, but the lack of any distinct advantages over other methods and human psychology are not likely to make this anything other than a rare occurrence. Finally, the scientists might develop and the public endorse some sort of screening technique whereby we can sort the sex-determining sperm in cases where married couples show a history of sex-linked defects. In the same direction, the public might endorse a controversial eugenic method such as that used by Dr. Georgiano Jagielo at London's Guy Hospital. In cases where she suspects chromosome aberrations in the egg may have caused a deformed baby, Dr. Jagielo induces superovulation in the woman. The dozen or so eggs are then carefully examined and their chromosomes studied. Depending on the percentage of abnormal eggs, the patient can be counseled as to the odds for her having a normal child. Is any one of these programs likely to become a reality? I doubt it very much, but a combination and highly modified program might very well draw on several of them.

How soon before a nationwide genetic counseling program becomes a reality? How soon before a nation, our nation, tries to solve its population crisis and perhaps improve its genetic pool by licensing the right to reproduce? No one today can predict. But many of those who have studied the possibilities—sociologists, medical officials, public health specialists, politicians, economists, biologists, and psychologists—cautiously warn that *it may well be much sooner than we expect.*

Robert W. Stock, staff writer for the *New York Times,* has noted a gradual change in the climate of the Western world and particularly the United States since the 1930's when Hermann J. Muller first advocated a eugenics program involving sperm storage and selective breeding. We have passed the fanatic extremes of Facist eugenics and overcome to some extent our fear and repugnance for the term eugenics. Contraception and family limitation have become a commonplace reality, not just among the wealthy and college educated, but now even among the slum dweller and the welfare recipient. As Frederick Osborn points out in *The Future of Human Heredity:* "Heredity [and genetic counseling] clinics are the first eugenic proposals that have been adopted in a practical form and accepted by the public." They are run by scientists, based on scientific knowledge, and geared to helping parents to avoid defective children. This is a very basic form of negative eugenics, indirectly contributing to the improvement of human heredity, even though the term eugenics is avoided in any discussion of these programs.

This new climate brings proposals such as that frequently advocated by the Nobel Prize-winning physicist, William B. Shockley, of Stanford University, within the realm of the feasible and the foreseeable. Dr. Shockley is convinced that the people of the United States will accept certain eugenic laws *within a generation.* His proposal for a sweeping population control plan, reported by Dr. Frank J. Ayd, Jr. in the *Medical-Moral Newsletter,* involves five steps:

STEP 1. Convinced by mass educational programs that some limitation is desirable and necessary for our survival, the American public endorses by democratic vote a specific rate of population increase, say one-third of a percent per year so that the population will double in two centuries.

STEP 2. The Census Bureau then applies this rate to the present population and calculates that this rate means 2.2 children for each girl reaching maturity.

STEP 3. The public health agencies insure that every girl is temporarily sterilized by subcutaneous injection of a contraceptive time capsule as she enters puberty. The time capsule might be the small silicon sponge developed by Dr. Sheldon Segal, biomedical research director of the Population Council. The sponge provides a slow seepage of contraceptive hormones so that the woman remains sterile until the sponge is removed.

STEP 4. When a girl marries she is issued twenty-two deci-child certificates by either the Census Bureau or local public health agency. She can visit her doctor and have the sponge removed in exchange for ten deci-child certificates. When she delivers her child, the doctor implants a new sponge to render her sterile again. The same procedure can be used for a second child.

STEP 5. After two babies, a couple may decide to sell their remaining two deci-child certificates through any member of the New York Stock Exchange or try to buy eight more on the open market and have a third child. Women who decide not to marry, nuns, and married couples who decide not to have any children could also sell their certificates without affecting the rate of population increase.

Startling as this proposal sounds, some version of it may soon become a moral necessity in the United States and other countries as well. As Dr. Shockley told his audience:

> My main purpose in proposing this example of eugenics is to provoke you to search your own conscience. Are you thinking seriously about those questions most important for the humans who will live in the world predominately shaped by the decisions of your generation?

Such a program will indeed solve our population crisis, but it amounts to a standoff, even allowing for the minor improvements brought about in the human gene pool by genetic counseling. Writing in the *New York Times*, Robert Stock has pointed out that "under pressure of the population explosion, *selective breeding theories will inevitably gain adherents.*" This is becoming more evident with each passing year, particularly as the public becomes more conscious of the reasoning behind such proposals through the mass media. In discussing "Eugenic Policies and Proposals," Frederick Osborn makes some valuable observations:

> It may take some time to develop effective eugenic policies based on competent scientific studies of the factors which affect size of fam-

ily. But meantime there is much that can be done along the lines indicated by our present knowledge and experience. We know, for instance, that in many large groups in the United States people tend to a rather uniform size of family. In the period after World War II it has been the fashion among high school and college graduates to have three or four children. During the great Depression of the thirties the fashion for high school and college people was for one or two children. In both periods those who went outside the number of children usual at the time were often made to feel that their behavior was odd, perhaps even anti-social. Such a standardization of size of family at whatever level is not compatible with a eugenic form of society. In order to provide a distribution of births favorable to eugenic improvement, there must be considerable variations in size of family within every group. This requires a public opinion with sufficient discrimination to appreciate larger families among responsible parents and to be impatient of irresponsible parents who bring children into a world in which they will get little care. It is not a matter of giving advice. It is rather a matter of the attitudes of those who help form public opinion in matters affecting the family. Doctors, ministers, and priests are close to the family, and their attitudes will have an important bearing on the many people they come in contact with. For people on relief or otherwise in contact with social agencies, a discriminating attitude on the part of community social workers together with advice on the availability of birth control should have a substantial effect. At every level of society there are responsible couples who desire children and should be encouraged and helped to have the children they can care for. Other couples who are irresponsible in regard to childbearing, whether wealthy divorcées or persons on relief, should feel the weight of an adverse public opinion, instead of the favorable attitudes which now too often accompany their childbearing. Community leaders of every kind should encourage this kind of wise discrimination.

Eugenics is much concerned with the selection of mates. Assortative mating, the mating of like with like, brings together constellations of genes alike for specialized qualities and concentrates them in family lines. . . .

As more is learned about the conditions that determine size of family among different kinds of people, new ways will be found to modify our social institutions for the specific purpose of increasing the proportion of children born to those who achieve most within the limits of their particular environments, and at the same time decreasing the proportion of children born to those with least achievement in their environment. Changes of this sort will be more easily accepted because they increase the proportion of children brought up in the more responsible homes in every environment.

The vast majority of politicians, economists, and demographers argue that the prime obstacle to an effective population control in the Western world is the lack of counseling and inexpensive contraceptives, particularly among the poor and less educated segments of the population. This is sheer nonsense according to a British scientist, at least in the United States. Dr. Peter J. Smith, a lecturer at the University of Liverpool, maintains that the chief problem is the excessive desire for children. Citing Gallup polls for the last quarter-century, Smith claims that the median number of children considered ideal by the average non-Catholic American couple has always been above two and that despite the differences in the various economic groups "on the average, all parents desire more children than the number required to maintain the population equilibrium." In an article for *New Society*, a British journal, Smith maintains that there can be no really effective population control until Americans change their attitudes as a society toward procreation. As things stand now, social and institutional pressures tend to stigmatize the childless couple and the single person. Married couples benefit much more from tax deductions than can any single person, primarily since our federal and state taxation laws are geared toward a family economy and culture. The elimination of tax advantages for big families offers a very practical and feasible first step toward reforming society's attitudes on this point, according to Smith.

Another step in this direction was taken by the executive branch of the American government in the late fall of 1969. At that time a leading representative of the Department of Health, Education and Welfare announced that federal- and state-employed social workers could urge a woman receiving welfare benefits to use some form of contraceptive and family planning. Prior to this statement, social workers were not permitted to recommend the use of contraceptives except when specifically asked about them by the recipient. An even more forthright statement came from Robert H. Finch, Secretary of Health, Education and Welfare, in late February of 1970. In an opening address to a conference in Washington on the environment and overpopulation, Secretary Finch was asked what people could do now on a voluntary basis to improve the environment. "I would begin," he responded, "with recommending that they start with two children." He also suggested that the federal government might have to offer "disincentives" to discourage parents from having more than two children per family though he did not spell out what these might be in any detail. On the international level, several indirect attempts have been made by the federal government to attach conditions favoring a national birth control policy for underdeveloped nations receiving foreign aid from the United States.

From these steps is it too improbable to foresee the day, perhaps within the next decade or two, when a family will be given tax exemptions for its first two children, none for a third child, and a tax penalty for any more than three? In the 1970's we may not accept "deci-child certificates," but indirect licensing of reproduction through tax measures may not prove so unlikely. For thousands of years man's psychology and society have been based on a high fertility necessary to overcome a high mortality rate. Medical progress has now made that notion obsolete, and a major reeducation is in order if man is to survive.

In his survey of genetic counseling in the United States, Robert Stock sketches a scene from the year 1975 when Mark and Linda are born. Before they leave the hopsital they are tested for a variety of known defective genes, as are all children born in this new world. Though they appear perfectly normal both babies carry certain recessive genes for defects and these are duly recorded on their "Universal Health Card" by the computer center in Washington, D.C. Later they meet, fall in love, and decide to marry. When they apply for a marriage license, their cards are called for and compared. Depending on the laws then in effect, they might be forbidden to marry. More likely Linda would be required to make use of artificial insemination if the risk was too high for a defective child, or undergo amniocentesis and other prenatal tests with a therapeutic abortion if no other remedy were available should her child be deformed. Radical as those proposals may sound as an infringement of innate human freedoms, the reactions to them today are not nearly as negative as they were to similar proposals made by Julian Huxley and Hermann J. Muller in the thirties. Recently, two renowned Nobelists, the British physiologist Peter Medawar and the American chemist and pacifist Linus Pauling, have suggested that two people who carry the same recessive defects, such as sickle-cell anemia, should be forbidden to marry. "I have advised, not entirely joking," Pauling admits, "that individuals should have tattooed on their foreheads symbols for the defective genes they carry. . . ."

However disturbing and even repugnant we may find these and similar proposals today, three trends seem fairly indisputable for genetic engineering and counseling in the days to come. First, it seems quite evident that the growing breed of genetic counselors will play an increasing and very crucial role in the mass reeducation of people everywhere, both in encouraging the limitation of our procreation and in directing it into selected pathways. Second, as our birth rate drops, each baby is going to become more and more unique as an individual who must be protected and nourished in every way possible from conception on. More effort can and will be devoted to each baby because there are fewer of them today, and because there will be even fewer tomorrow. And finally, psychologically and emotionally, we of the older generation, and by that I mean everyone over fifteen, must start preparing ourselves for the imminent revolution that we will observe in society and in the families of our children and grandchildren. Dr. German pointed out some time back that the "admirable advances" in teaching high school and elementary school biology have prepared today's youngsters to accept and appreciate the value and need for genetic counseling and planning. Though our educational processes have a long way to go, these same admirable advances have conditioned our children to accept a revolution far more serious and extensive for mankind and society than most of us imagine.

SIX

HUMAN
SEXUALITY
IN PROCESS
Where Do We
Go from Here?

THE PILL! How often have we seen that product of quiet biological research ballyhooed in opalescent and reverential headlines? Despite the fact that there are more than two dozen quite
different types of sequential and combination contraceptive pills
on the market today and despite the fact that far safer, more
sophisticated second and third generations of "the Pill" are in
the wings, "the Pill" still remains a mythic symbol, a terse and
majestic tribute to the new embryology and its exploding role in
society. But this reverential usage indicates something far deeper
than the mere liberation of women from the fear of pregnancy
and unwanted offspring. "The Pill" is also, as many people subconsciously suspect, the symbolic symptom of a major revolution
already under way.

In discussing the future of human sexuality, Marshall McLuhan
and George B. Leonard pointed out how subtly and subconsciously new ideals, roles, and conceptions of the human male

and female are being triggered by the technology of the Pill. The contraceptive Pill, they maintain, matches the explosive power of the nuclear bomb and like it can wipe out in an instant all social boundaries related to sexual behavior. The Pill, according to Mc-Luhan and Leonard, turns woman into a bomb that creates a new fragmentation in human behavior, blurring the old boundaries between male and female by making it possible for woman to experience the same freedom in sexual behavior as man.

Ashley Montagu recently argued that the Pill "*is* a major revolutionary development, probably to be ranked among the half dozen or so major innovations in man's two or more million years of history." In his opening paragraph to *Sex, Man, and Society,* Montagu explained that in its effects "the pill ranks in importance with the discovery of fire, the creation and employment of tools, the development of hunting, the invention of agriculture, the development of urbanism, scientific medicine, and the release and control of nuclear energy."

Montagu's equation is to my mind as indisputable as it is obvious. However, it is strange that Montagu does not reinforce his argument in *Sex, Man, and Society* by exploring, or at least mentioning, the rapidly emerging technology of human reproduction of which the Pill is but a minor, though symbolic facet. The social implications and ethical revolution of the Pill are *minuscule* when compared with the social and ethical impact of the technologies we have reviewed in the preceding chapters. In this context, the Pill is a nonproblem because it is already passé in man's exploding technology of reproduction.

More than a decade ago the noted German theologian, Romano Guardini, spoke with apprehension of "the end of the modern world" as symbolized in the appearance of "technological man." Despite the lingering presence of poverty and illiteracy in major segments of the population, mankind as a whole, to use Raymond Aron's phrase, is entering an "epoch of universal technology." The survival advantages provided by a technological culture have already doomed to the role of peripheral backwashes or lingering anachronisms such human cultures as the Australian aborigine, the Amish, and the southern rural culture of the United States.* The reality of past everyday customs will naturally continue into the era of the new man, and the world is likely to remain for some time to come a confusing crazy quilt of the old and the new. One vital factor, however, distinguishes the present situation from past human experiences. In the past, geographic isolation allowed men to adjust rather comfortably to new discoveries and inventions. Today mass communications and travel have demolished the shelter of isolation so that countless traditional customs and social patterns are guaranteed an accelerating death.

The emergence of technological man is a fact we must face. The question is, How do we cope most effectively and wisely with his reality? Do we have any resources or guidelines that arm us for an encounter with what appears at first sight to be a threatening and overwhelming peril to humanity and man's humaneness? Teilhard de Chardin remarked with repeated insistence that the current of our past history holds *the key* for any understanding of our future. Nowhere do I find this insight more helpful than in studying the emergence of new patterns in human society and the changing relationship of men and women. Distinct trends in this phase of cultural evolution are particularly evident in the short span of the last two centuries on the North American continent. But they are also apparent in the tedious groping emergence of Western man's sexual consciousness over the past two thousand years and in the several million years of sexual evolution among the mammals and primates prior to man's appearance.

The frontier farmer was up with the sun and out in the fields for ten, twelve or more hours of hard manual labor, and this for some 320 days of the year. His travels were limited to a weekly or monthly visit to the nearby town for provisions and an occasional social outing. Sunday brought the whole family to town for church services while the evenings were invariably

* Commenting on this persistence of old orders with their institutions, ideals, and ruling classes, Victor Ferkiss has noted that "when one speaks of the advent of the new man on the existential level, one need not assume his [immediate] triumph on the social level."

spent around the family hearth. Even in the largest towns this pattern was repeated with only slight and nonessential variations. Hard manual labor in the factories replaced the plow, but the structure of society was merely compressed from the open ranges to the tighter network of the neighborhood block with its individual homes or tenements. Generally, although this was even then in the state of flux, the place of women was set in the home with the children and few dared wander beyond that boundary. Contact between men and women beyond the broad family circle was generally restricted to surface amenities and business. Outside of courtship, it seldom involved deep personal contacts with those of the other sex.

Today's urban Americans, despite allowable differences in detail owing to economic and social dissimilarities, inhabit another world altogether. Our mobility among the varied fields of endeavor associated with modern industry has shattered the sheltering fences of our grandparents' farms. Often mental and sedentary rather than manual, our work schedule may call for a five-day, eight-hour-a-day week and some 250 workdays a year. Occasional overtime or moonlighting with a second job may add to this, but men today work fewer hours than ever before in human history. Instead of working within earshot of the family hearth where his spouse could bring him lunch, the modern male frequently commutes many miles a day between home and work. Coupled with these changes the varied world of urban life has managed to throw the modern male into a continuous procession of encounters with women on a multilevel personal basis: secretaries, co-workers, and associates as well as casual friends.

This same evolution holds true for the modern woman. As urbanization took hold women frequently had to leave the hearth to work beside men in supporting their families. Two World Wars accelerated this shift which pervades much of our culture today. Women are now far better educated than ever before and often have a profession, full or part time, of their own to fill their time. Urban public transportation and the suburban second automobile allow her full advantage of newfound leisure. With her children occupied from nursery school through college, from the age of three or four to their twenties, leisure and variety have become a reality for woman. The automatic dish washer, the clothes washer and drier, whether at home or the automat, have turned the Monday wash, Tuesday ironing, Wednesday cleaning, Thursday bridge, and Friday shopping cycle into a chaos of activities. All the conveniences of modern life plus the demands of supporting her husband in his work-oriented social life have demolished the private sheltered world of woman-wife-mother.

In this rapidly evolving culture the basic social unit of the family was bound to be modified, first geographically and then technologically, by an extension of the trend toward what Margaret Mead likes to call the "nuclear family" of husband, wife, and one or two children. Near relatives scattered over the countryside and even across the oceans. Instead of moving next door or down the block, children began moving clear across the continent when they left father and mother to take a wife. Friends, instead of being around the corner or clustered within a few city blocks of each other, now are sprinkled across several states. Like a radioactive atom the human family is shooting off its particles in many directions. But the tendency toward a true nuclear family culture is not unresisted, even by modern technology. The ubiquitous interstate highways with their air-conditioned passengers, the jet airports, the often muddled postage stamp and telephone cable, despite their defects and limitations, nevertheless provide communications links extending the human senses so that the fragmentation process is retarded and highly modified. This phenomenon holds true on all levels of modern society.

In the broad context of this social evolution, two specific byproducts of technology have played an important role, particularly in terms of the ethics associated with family life and the interpersonal relations of men and women. A generation ago parents viewed with great anxiety the privacy and mobility provided for their sons and daughters by the automobile. With the advent of the Ford assembly line, practically every teen-age couple had access to a convenient and portable bedroom. Today we can add to this, for both the affluent and not-so-affluent American, thousands of motels on nearly every highway of our nation which offer the privacy of a home away from home.

This trend has only been accentuated and accelerated by the

Pill and similar contraceptive devices. However, unlike the other developments mentioned above, the Pill has made a revolutionary contribution to human evolution because it has brought mankind, if not yet in its psyche, then at least in concrete practice, to the brink of what geneticist H. Bentley Glass bluntly labeled "a total severance of human sexual life from the reproductive function." While mankind for the moment still relies on sexual intercourse for procreation, he nevertheless has gained almost complete control over its reproductive aspect. The individual adult today is morally responsible for *two distinct decisions,* the decision to reproduce and the decision to enter into the total dialogue of sexual intercourse. Men and women today are responsible for the separable, if not yet always separated, activities of sexual communion and procreation. Through the pill both these activities have finally become *truly human.* Mankind now exercises the choice between irresponsible and uncontrolled reproduction and responsible mature parenthood. Men and women alike now face the freedom to enjoy sexual relations and intercourse either as a hedonistic egotistic pleasure or as loving interpersonal dialogue-communion.

All of these trends in human society over the past two centuries have forced a radical change in woman's attitudes toward herself and her sexual (genital) activities. But they have also brought about a complementary, perhaps even more radical change for men in their attitudes toward women as persons, toward themselves as males, and toward the language of the body.

This discussion leads us further back in the history of evolution in our search for some key to sexual man's future. Certain instructive developments and trends are apparent in the evolution of sexuality among man's early primate ancestors and in fact among the more remote mammals. It was from observations among these subhuman animals that Western man developed that radically false equation for man which essentially links sexual intercourse with reproduction, and, in the extreme, only justifies sexual intercourse when it is within a procreative context and function.

There are some substantial differences in the simple biology behind the sexual union of the lower mammals and the same activity exercised by man and the higher primates. Prime among these is the fact that two different cycles regulate the matings in these two groups. Among the lower mammals an estrous cycle brings the female into sexual heat only when she is ready to conceive. This may be once a year as with the deer, more than once in a particular season, or several times during the year as with many domesticated animals, but outside these well-defined short periods the female and male are generally indifferent to each other. Just as the female only ovulates during heat or in a reflex to mating as with the rabbit, so the male produces sperm only during the same short season. Thus mating seldom occurs except when everything is geared to assure conception. Among certain monkeys, the ape and man, a quite different sexual cycle appears, a product of recent evolution and natural selection. Especially in women but also in her close biological relatives, the monthly menstrual cycle has little if any direct relationship with the female's sexual interest or inclination to mate. The female may be fertile for only three or four days in her monthly cycle, but her willingness to enter into sexual dialogue with a male is hardly so circumscribed. While ovulation occurs on a regular cycle, spermatogenesis is continuous from adolescence to old age. Man and his near "cousins" (both male and female) are freed of the physiological and instinctual compulsion to mate only when conception is assured. An estrous deer spends one-tenth of its life dependent on its mother, and about 85 percent of its life span devoted to reproduction. The human child, on the other hand, is dependent on its mother for a quarter of its life and women devote less than a third of their lives to procreation and child rearing. Even in the most fertile marriage where no contraceptives are used, the marital union is procreative well below 1 percent of the time.

The anthropoids and humans are the only sexual animals known to mate regularly during pregnancy, lactation, and after menopause, times when conception is either totally impossible or highly unlikely. Further decreasing the connection between intercourse and procreation for man is the fact that natural selection has failed to provide women with any means of storing sperm when mating occurs before ovulation. Such structures have evolved to assure and increase fertility in many insects, bats, and

even cows. In quite an opposite trend, the reproductive system
of the human female has evolved a barrier at the opening of the
womb and chemical secretions in both the womb and vagina
quite lethal to sperm except during the limited time of ovulation.
Further highlighting this trend toward the separation of sexual
union from procreation is the peak in sexual desire which for
many, if not most women comes just before or after the menstrual
period when conception is most unlikely.

Converging evidence of this trend can be found in the science
of ethology or primate behavior. Here the anthropoid apes, the
monkeys, gibbons, chimpanzees, and baboons provide an instructive variety of steps in the evolution of the primate behavior of
which man is the latest edition. A prime characteristic and advantage of the primates is their superior brain development. Because of this, primate and human babies are born at a stage of
development and dependence very similar to that of the marsu-
pial opossum and kangaroo. Survival of the helpless newborn,
and hence of the species, depends on a close and devoted relationship existing between the mother and offspring. This relationship, however, opens the mother to attacks from the outside and
difficulties in obtaining food for herself and child. This has been
remedied through natural selection by a sexual system which
tends to keep the male around most of the time. A female who
is almost always open to sexual activity is much more attractive
biologically speaking than one who comes into heat only once a
year. The result has been the emergence of a closely knit continual association of mates, often within a monogamous pairing,
but also within a small band or troop. Here lies the start, biologically speaking, of the human family and the groping, tedious
transition from mere procreative compulsive instinct to the sexual communion of male and female in a community of interpersonal love, affection, and in-depth sharing.

Linking this biological evolution of mammalian sexuality with
the cultural evolution of the past two centuries is a period of perhaps two million years when man's social evolution interacted
with his emerging awareness and deepening understanding of
his own sexuality to further reinforce the trends noted above.
Western man's evolving image of himself as a sexual person has

been deeply influenced by four major social or cultural currents, the Hebrew, Greek, Roman, and Christian.

In many respects the oldest of these influences has also been the most vital and wholesome. The oriental tradition of the Hebrews as recorded in the Old Testament and the Talmud had its origins somewhere in the Middle East some three or four thousand years ago. Its oldest oral traditions eventually were placed in written form with the editing of Genesis, which brought together at least two distinct though complementary images of the male-female relationship. The older of these, the Yahwist tradition, can be traced back to the tenth century B.C., most likely to the reign of Solomon. It deals with human sexuality, as it deals with the wholeness of the human embodied person, in terms of psychological and fully human companionship, recalling that "it is not good for man to be alone." Sexual intercourse is viewed not as a mere physical activity, but as an expression of love, a means of becoming one flesh, an expression of human communion as well as of the human need for such communion. The term most often used by this tradition to describe sexual intercourse was not euphemistic, but rather a penetrating terse summary of what this should be, "knowing." The second Hebraic tradition, that of the priests, was probably put down in written form after the Babylonian captivity ended in 538 B.C. Its focus is on human sexuality as a biological reality imposing the duty to "increase and multiply" common to all animals. But even here man's sexuality is placed on a somewhat different level from that of the animals, for God creates *both* man and woman "in his own image and likeness, male and female." Psychological personal communion and procreation were for the Hebrews two complementary functions of human sexuality. But even here, at a time when man had absolutely no control over conception as the unpredictable outcome of some acts of sexual union, the prime emphasis was on sexual intercourse as a deep interpersonal mode of communion and dialogue.

One major aspect of the emerging monotheism of the Hebrews was their refusal to grant to Yahweh a female consort as all the other religions of the East had done with their gods. While the other early religions sacralized human sexuality in their religious myths, the Hebrews turned to a prehistoric human couple, Adam and Eve, for their model of human sexuality. The pagan could turn to his religious myths, the stories of Apollo, Venus, Aphrodite, and others, for guidelines on their sexual images and behavior. Not so the Hebrew. Man has always had to grapple with "the difficulty of freeing [our] thought from divine archetypes and of effectively desacralizing and demythologizing sexuality. . . . The humanization of sexuality," in Joseph Blenkinsopp's appraisal of *Sexuality and the Christian Tradition,* "is an essential aspect, perhaps *the* essential aspect, of growth towards a genuine personhood."

Blenkinsopp, expanding on a point made earlier by Jean Daniélou, has raised a valuable point for our understanding of human sexuality when he points out "the liberation of the sex act [in Hebrew thought] from this kind of [pagan fertility] mysticism and its location in the context of an on-going dialogue, a mutual self-giving between man and woman." The Hebrews had no word for sexual intercourse, nor even a word for sex. They spoke of man and woman not as isolatable and distinct beings, but rather as persons existing in and growing out of interpersonal dialogue. The individuality and uniqueness then of the male and female came from "a calling of each other into fuller being, a mutual discovery of the one in the other." Blenkinsopp goes on to explain that the sexual man

> exists as response both to the call of being [from the Creator] and the word which is addressed to him by the other [human person]. Relational existence means a continual calling into being [creation] of the other as I, in my turn, am continually called into being by him. This opens the way to overcoming the dichotomizing of sex as either demonic or institutionalized [in marriage] and to understanding it in the categories of communication, language, address and response.

In certain ways the Hebrew woman was better off than her counterpart in other Semitic societies. In theory and more often than not in practice the monogamous marriage was the ideal. Yet in the patriarchal pastoral orientation of the Hebrew people, marriages were arranged as family alliances by the parents. Until a girl reached roughly the age of twelve, she remained the ward of her father and was often viewed as a useful if respected servant in his household. Her seclusion gave some guarantee of her remaining a virgin till marriage, a matter of considerable concern for both her father in terms of the expected dowry and for her prospective husband in terms of his heirs and firstborn son. Early marriages were favored as an effective way to protect the girl's virginity and also assure many children for her husband. Woman's place was clearly in the home, where both her dignity and influence depended on her producing many heirs. If she should prove to be sterile, her husband could divorce her. But even when she produced heirs, her husband might decide to take another, younger wife into the household as she grew older. Seldom did marriage bring anything more for the wife in the way of personal rights and development other than the passage from being a ward in her father's house to a similar position as a "child-wife" of a man often much older than herself. Love between a man and woman seldom entered into consideration in what was basically a matter of social and economic agreement by two fathers.

The fact that the Hebrew people constituted a very small nation, often exiled and under attack, diverted them from the ideals contained in Genesis. Yet despite the frequent emphasis on fertility and the common treatment of women as objects of an anti-feminine pessimism reflected in a definite subordinate status and social limitations, Hebrew women did occasionally break through the mold to share the ideal image of male-female relationship portrayed in Genesis. When they did, the result was magnificent: witness the beauty of Sarah, Rebecca, Rachel, Deborah, and others.

In this patriarchal culture with its practical emphasis on woman's fertility, the double moral standard flourished. Men, for instance, could institute divorce proceedings on a variety of grounds. Not so the wife, though she might ask her husband to divorce her. Premarital and extramarital intercourse were formally forbidden, but the risk and penalities were far graver for the woman than for the man, frequently involving the death penalty for the former and a fine for the latter. Brides were frequently required to prove their virginity; not so the husbands. Thus, despite a major advance toward a more human relationship between the sexes on the plane of ideals, a heavy thread of male dominance and female subjugation and seclusion winds throughout the Hebrew culture.

The second major current to mold our present awareness of man's sexual nature came from the pre-Christian Greek world of Socrates, Xenophon, Plutarch and the democratic traditions of Sparta. In setting out the ideal structure and functioning of his *Republic*, Plato later argued for an equality among men and women in the ruler class. He likewise posed a strong case for communal marriages in which parenthood would be shared and all the children would be considered brothers and sisters to each other. The political and social factions fostered by individual family loyalties would then be broken. An added advantage would be the possibility of the rulers secretly rigging for eugenic purposes the communal marriages which were to be decided by a lottery.

The wife of the average Greek of some wealth was for the most part excluded from public life. Her place was the "women's quarter" of the house where, carefully guarded from men, she supervised the household, educated her children, and listened to the superior wisdom of her husband. The wealthier Greek male was accustomed to the charming company of the hetaerai, the higher class mistresses, who were well educated and often much more cultured than the wives as well as trained in pleasing the male. In some cases the hetaerai had the added prestige of being associated with religious sects where prostitution was a vital part of the religious practices. Other avenues to fame and influence were open to the hetaerai: Pericles eventually married Aspasia; Thais was the influential mistress of Alexander the Great; and Phyrne modeled for Praxiteles' famous sculpture of Aphrodite, the goddess of love. Generally Greek women had three choices: to become a wife, take care of a household, and bear children, to accept the role of concubine, or become a hetaera and entertain the husbands of other women. Choosing the first life severely restricted her contact with men other than her husband because of the prevailing double standard. As Demosthenes expressed it so well, "Man has the hetaera for erotic enjoyment, concubines for daily use and wives of equal rank to bring up children and to be faithful housewives." As with the Hebrews, marriages were arranged by parents or clans and love seldom entered the picture either before or after the wedding.

The culture of Rome also made a basic contribution to our present image of human sexuality. The early position of the Roman wife varied little from that found in Greece except that her competition from the courtesans was not nearly as strong. Neither highly educated nor cultured, the Roman counterparts of the hetaerai occupied a much lower position in Roman society. Furthermore, the Roman wife could attend banquets with her husband and appear in public with him, something strictly forbidden to the Greek woman. Idealizing their women, the Romans placed great value on female virginity, as is common wherever the double standard of morality holds sway. Less aesthetically inclined and coarser in their tastes than the Greeks, the Romans approached sexual relations, marital, premarital, and extramarital, with an earthy practicality.

The Punic Wars, from 264 to 146 B.C., brought important changes in Roman society. With increasing wealth, urbanization, the rise of a leisure class, the importation of countless slaves and the absence of many Roman men from home for long periods of time, women began to gain both social status and legal rights. The husband's power of life and death over his wife was revoked. A woman could inherit property and even relax in the presence of her husband when he sat. Common law marriages were strengthened and divorce laws equalized to the point where mutual consent was accepted. As woman's participation in society increased, her role as chattel object necessarily decreased. Though weakened, the double standard nevertheless persisted. As the great orator Cato the Younger remarked, "If you find your wife in adultery, you may kill her freely without a trial. But if you commit adultery or if another commits adultery with you, she has no right to raise a finger against you." While Greek law occasionally encouraged limiting the family size, the Romans viewed this as a heinous crime.

During its first three centuries, before the reign of Constantine, Christians had very little influence on public life. They were a poverty-ridden, despised, persecuted minority with radical views on many subjects. From the beginning Christians opposed the new position of woman, the liberal divorce laws, and other changes in society wrought by the Punic Wars. Despite the fact that they ordained women as deaconesses, accepted them as prophets in the church, and allowed married priests for at least four centuries, two major forces brought into Christianity a very strong bias against women, marriage, and sexuality. The first of these was internal, the belief in the imminent end of this world and the return of Christ. If this world was about to end and the Second Coming was at hand, then it was no time for Christians to indulge in marriage and start a family. The other influence, while extraneous to Christianity itself, has nevertheless been a major cause of the distortion which viewed women as somehow the embodiment of evil because of their sexuality and the temptation they posed for men. This philosophical view arose with the Persians, Greek and other Eastern mentalities, which tended to view the world in terms of black and white, good and evil. Such, for instance, was the cult of Mithras, the Persian god of light and truth, which was Christianity's strongest rival in the days of imperial Rome. Among the Greeks this dualistic splitting of the world into two realms was applied to man by Plato and his followers. Man was composed of two distinct substances, one spiritual and godlike, the other demonic and evil. In Platonic thought man, in essence a godlike spirit, was imprisoned in a material body until death freed him to live the true life with the gods. Introduced into Christianity by Augustine and Plotinus in the third century, this pagan view of man soon overwhelmed the more wholesome, human approach of the Hebrews, who viewed man simply as a living person and refused to speak of him in terms of body and soul. This dualistic pattern of thought soon permeated man's image of the world. On one hand there was a sacred timeless realm of ritual worship and the spirit; on the other, the profane, temporal world of matter, flesh, and sexuality. Such a view left no room for the erotic in man's spiritual life. Dominating the Christian world for the next thousand years, this gnostic view combined with the expectation of Christ's imminent second coming to exalt virginity as *the* ideal for all men and women. This view appears in the almost extreme reluctance with which theologians of the Middle Ages dealt with marriage as a way of life compatible with man's spiritual life and goal, as well as in the strong disdain in Christian thought for anything touching on physical sexuality.

In apostolic times both Peter and Paul endorsed the concept of the child-wife whose rights were almost nonexistent and who should be "used" considerately by her husband. A wife's submission and obedience in both family and society was granted both scriptural and philosophical foundations. Where in apostolic days Paul pointed out that it was better for a man to avoid any physical contact with women, three centuries later the arch misogynist Jerome went much further to claim that it was sinful for any man to touch a woman, even his own wife. Methodius argued that women had become by environment what nature had made them, "carnal and sensuous," "the irrational half of the human race." If Jerome was correct in viewing woman as "the devil's gateway" and a "scorpion's dart," then the danger to man and his spiritual life was obvious. Slow to understand, unstable, naïve, mentally weak, women definitely needed the strong authority and wise direction of a good husband, if they did not enjoy this from a spiritual director as a cloistered religious. When John Damascene spoke of women as "a sicked she-ass," a "hideous tapeworm," and the "outpost of hell" in his famous sermons, it was only natural for Pope Saint Gregory the Great to conclude that the proper "use" of woman was twofold: prostitution or motherhood. Even one's own wife became a harlot when she was used for any other reason than to have a child, or if she were loved too ardently.

This view of woman naturally had its impact on the common view of marriage which was seen as a second-rate vocation for Christians. If one simply could not overcome the sexual drive and abstain from sexual intercourse, then it might be better to marry than burn in hellfire hereafter for sinful indulgences of one's uncontrolled passions. The sexual passion was *an impersonal and explosive force* which must be resisted because it was so opposed to man's spiritual growth. In a church where most priests were married, Ambrose nevertheless argued that the marital union was so animalistic and defiling that no priest should ever marry, and if he were already married then he should live with his wife in the continency of a brother-sister relationship. Jerome forbade married couples from receiving the Eucharist for several days after they had performed this "bestial" act and informed husbands that they actually insulted their wives by having relations with them. For fifteen hundred years, until the Council of Trent, theologians debated fiercely whether marriage could be considered a sacrament when it involved such obvious bestiality. Even when the sacramental character of marriage was accepted, total or at least partial abstinence from sexual intercourse was strongly recommended for all married couples, particularly by the Irish monks. Nonsexual, celibate marriages and the state of virginity were extolled, for unlike the Hebrews who openly praised sexual love and the sensual in the "Song of Songs," the Christian Church has seldom found even a small niche for the erotic and the sexual in its theology.

Despite all this deprecation of the feminine and the sexual in human life, actual daily life in the Western world was more moderate and sensible. Though it continued to crop up in everyday life, the double moral standard was at least rejected in theory by Church authorities. The Church finally pressured the civil authorities to outlaw civil divorce in the fifth century after considerable debate. The basis of their argument, however, offers an illuminating commentary on the clerical mentality of the day. Divorce was ruled out not for any biblical reason but rather on the assumption that divorced men were likely to remarry out of lust rather than love.

As the Christian culture matured and the Second Coming receded further from view, new appraisals of the male-female relationship emerged in the early Middle Ages. On the surface the concept of courtly or romantic love appears to be a more mature and human understanding of sexuality because it brought to the fore woman as a person. But beneath the surface there is the strong gnostic and dualistic influence of the Arabian philosophers who introduced Aristotle to the western world at that time. The woman placed before man as his ideal by the troubadours of Languedoc was very much an angelic creature, an asexual, and hence ultimately dehumanized, person. The divine Beatrice whose beauty, wisdom, and noble character Dante acclaimed was such a lady, another man's wife, ethereal, beautiful beyond belief, radiant, almost beyond the reach of mere mortal man. Troubadours and knights composed their ballads and serenaded the divine perfections of their noble ladies, in whose name it was glory to give one's life in mortal combat.

But Camelot was more than a romantic idyll. It was in some respects a reality, but it was also a never-never land, a unique world in which the *inhuman* dualism of gnosticism and the Albigensians could revive itself beneath the guise of a humanizing love. The troubadour's beloved lady, like Lancelot's beloved Guinevere, was almost invariably the aristocratic wife of another man. As in Camelot, the courtly love affair blossomed in a castle filled with bachelor knights and troubadours, servants and wenches, and one or perhaps a few aristocratic ladies. Thus was the unattainable rhapsodized, but not merely because it was unattainable. People in those days generally believed that love and marriage simply would not mix with any success. Love did not enter into their picture of marriage. This dualism of the evil body and the divine spirit of man came to the surface in another way also. To endure, love must remain free of both marriage and sexual intercourse. Hence, to consummate one's love of a lady with sexual intercourse meant its death as witness the stormy romance of Heloise and Abelard.

In the early days of courtly love, very little if any sexual expression was permitted. Everything was quite platonic, as when Lancelot visited Guinevere in her chambers and nobly listened to her confess "I loved you once in silence." But at least tenderness and affection were introduced into the relationships between men and women. By the fourteenth century the platonic character of courtly love was eroding, and a lady might reward her lover's devotion by allowing him to spend the night with her provided he swore continence. This new custom, however, was very unstable and by the middle of the seventeenth century extramarital intercourse with a lover was formally accepted by the upper classes of society. The uneducated peasants, of course, had their own views on the relationship that should bind men and women.

By the seventeenth century romantic love captured the fancy of the new middle-class merchants, although any formal acceptance of adultery as a reward for love caused many aristocrats and most of the newly emerged, conservative middle class serious concern. The solution came by way of a shift in the person of the beloved maiden, from the married woman to the single girl. With parents still very much in charge of arranging marriages for their children and guardians quite watchful over their maiden daughters, this limited the love-theme to engaged couples who might be allowed some privacy and acquaintance before they married. Inevitably this move led to the young people themselves asserting their rights to choose their own mates and marry for love. By the eighteenth century, particularly in England and frontier America, this revolution was complete, at least among the poorer classes.

The Industrial Revolution and the massive process of urbanization gave added impetus to this evolution, but a major contribution came from the women themselves in the form of the feminist movement. With the ground broken in France by Voltaire, in England by Mary Astell, and by Benjamin Franklin in the States, Mary Wollstonecraft's *Vindication of the Rights of Women* could trigger the first "Woman's Rights Convention" in Seneca Falls, New York, in 1848. Despite Queen Victoria's plea for women to remain at home where they belonged, countless pamphlets, including the potent *Subjection of Women* by John Stuart Mill, added fuel to the revolution. In 1869 the women of Wyoming won the right to vote and the movement sidetracked into the prohibitionist campaign of saloon-busting Carry Nation.

The pioneers of the feminist movement introduced many social changes, some of them of small if prophetic impact at the moment. Lucy Stone, who with Julia Ward Howe had formed a conservative splinter from the more radical Susan B. Anthony and Elizabeth Stanton, refused to give up her maiden name when she married. Her husband, Henry Blackwell, was probably not too disturbed at this feminine rebellion since his two sisters became the first women doctors in America and his sister-in-law the first woman minister in the States. Deeper changes came in the field of education where, despite the warning of Harvard's President Eliot that they were not equipped by nature for the strenuous academic life, women began to pursue higher learning. In 1910 eleven thousand women graduated from college; today that figure is over three and a half million college graduates each year. With their first tentative steps into the labor market as secretaries and seamstresses, women began to mold a new world for themselves. Today roughly twenty million American women, half of them married, are employed outside the home.

After thousands of years during which the female half of the human race labored under the stigma of being "defective, ill-formed males," to use the description of Thomas Aquinas, and therefore being doubtful members of the human race, women finally began to emerge as human persons in their own right. With an education and economic independence they could no longer be merely tolerated because of their procreative and maternal functions. Generation after generation of women had shared a servitude not unlike that of their underaged children except that their legal, economic, cultural, and psychological dependence on the male lasted throughout their lives. Then, within the space of a single century, a major transformation occurred in what Teilhard de Chardin has wisely termed the *personalization* of women. Women became humans. Though the social gap between men and women is not yet resolved and countless inequalities continue to exist in education, industry, and society, women have in fact begun to take their rightful place as mature, independent persons within the human race. The labor pains of this transformation, far from lessening today, are likely to become more acute, particularly for men, once the impact of the new techniques of human reproduction we have surveyed here become widespread.

Brief as these sketches have been in summarizing the evolution of mammalian and human sexuality, they do provide us with a situating background against which the trends inherent in the new embryology are highlighted quite clearly.

Overshadowing all other trends within the new embryology is the fact that, after nearly two million years of evolution and only 5,000 years of civilization, man is finally coming to accept his sexuality as a dynamic process, coextensive with his development as a sexual person. This developmental process applies not only to man as an individual sexual person, but also to the human race as a society comprising countless generations of individual sexual persons. Mankind, like man, goes through an infancy, childhood, adolescence, and eventually should reach a stage of maturity.

In the past man's perennial attempts to analyze and conceptualize human sexuality have without variation started off from the premise that sexuality is some sort of epiphenomenon, accidentally tacked on to an unchangeable human nature. From Aristotle and Plato, to Paul, Augustine, Aquinas, and through the Middle Ages, on to such modern proponents as Gandhi and Norman O. Brown, author of *Life Against Death* and *Love's Body*, human sexuality has fallen victim to an unyielding determination to locate it somewhere—anywhere—outside the human self, outside the authentic "me" and that inner core of personhood which makes man distinctive. For this strain of thought, human sexuality is a nonessential, truly accidental appendage of human nature. Human sexuality has been for scholars of this bent far too explosive and unpredictable to permit its incorporation into any rational philosophy of man. It was likewise too pervasive yet nebulous for scholars to tolerate in their quest for an objective logical system of thought. The only comfortable way to deal with human sexuality was then to "reify" or objectify it as *something distinct from man's essence*. Thus arose those marvelous and alluring distortions, the perennial mystiques of the Eternal Feminine and the Eternal Masculine. Unchanging archetypes of a classic Platonic ideal world (and not in the Jungian sense), these objectified mystiques reduced human sexuality to a thing, an objective accident quite apart from the core of human nature.

In societies as resolutely antisexual as Christianity in the West and the Taoists, Hindi, and Buddhists in the East, these unchanging abstractions offer a comfortable if facile means of handling a very explosive subject one cannot emotionally or intellectually cope with. The problem of human sexuality basically has always been how to integrate this reality into our religious thought on man's spiritual progress. Christianity and many of the Eastern schools of thought have solved the dilemma by reducing human sexuality on one side to biological generation and on the other side erecting it into idealized mystiques of what a male or female should be in terms of behavior. The ideal good woman then became the quiet, shy, stay-at-home, nonintellectual, nonpassionate bearer of children, housekeeper and support of her husband. Any woman who sought to express her self-identity and personality by having a career outside the home, pursuing an education without marriage as a prime goal, or expressing an interest in sexual fulfillment could not be a "good woman" or a fit mother of children.

As long as human sexuality is viewed as something we *have*

rather than something we *are*, these abstractions can stand. But if there is one lesson modern psychology, from Freud, Jung, and Adler down to recent developments in group dynamics, has taught us it is that human sexuality is *what we are*, not what we have. As Abel Jeannière, author of *The Anthropology of Sex*, said so well, "sex is what man becomes." Human sexuality, like human nature, cannot be defined in the abstract. To speak of what it is to be masculine or feminine outside of the dynamic history of a person's relationship with other sexual persons is impossible. Human sexuality is, as Daniel Sullivan points out in the preface to Jeannière's book, "a mobile genesis coextensive with human genesis." It is "an aura and an all-encompassing factor as well as source of the human being." In this process context human sexuality is a reality, Jeannière assures us, "which cannot be described once and for all." *Human sexuality is coextensive with human personality.* It cannot be reduced to genital biology nor abstracted into a psychosexual mystique.

Psychologists have long tried to find an objective basis for a description of what it means to be masculine or feminine. In our American culture we might place in a masculine lineup such personality traits as strength, virility, bravery, aggressiveness, enterprise, mechanical interests, propensity for outdoor activities, and adventurousness. Among feminine traits we could list domesticity, sedentary activities, softness, tenderness, emotional affections, and sentimentality. But common sense and everyday experience make us very aware of the arbitrary nature of such listings. Naturally tests developed by the psychologists on the basis of the above lists reveal great ranges and variations in masculinity and feminity within both biological sexes. For instance, with their standard test of the masculine-feminine factors Professors Lewis Terman and Catherine Miles have found that college athletes, engineers and architects, and high school boys score very high on the masculine scale, with bankers, dentists, teachers, doctors, clerks, and construction workers about in the middle and policemen, firemen, journalists, artists, and clergymen more inclined toward the feminine side. In fact, women Ph.D.'s and doctors along with the average college women appeared more masculine than the policemen, journalists, and clergymen on the Terman-Miles "M-F" Tests. Much of this variation can undoubtedly be traced to social conditioning and education, a fact which again raises the issue of human sexuality as a process coextensive with the process of personality development.

Ashley Montagu maintains that "the baby is not born with a human nature, but with the capacity to become human." An egg is fertilized and cleaves. It has no personality, no sexuality save the information captured in the sugars, bases, and proteins of its microscopic chromosomes. Yet within hours two cells become four, and the geometric division continues until a week or so later the small mass of cells embeds itself in the lining of the womb. There it grows and develops, taking on ever so gradually a human form, learning to move, learning to hear and drink. With the contractions of the womb it is pushed into this world and pampered by its loving parents. Pink or blue booties are placed on its tiny feet and it begins in earnest the lifelong task of becoming a mature human person, masculine or feminine as the case may be. The umbilical cord has been cut and this embryonic person we cuddle has made a major step in a painful but also joyously creative process. Weaning will come as will the nursery school traumatic separation from home, adolescence, and adult life—but always a process of growth which can only come about when we are in dialogue with other people. Each of us becomes more deeply human and more self-creative as persons through interpersonal dialogue with the universe and with our fellow men. Our groping toward masculinity or femininity likewise comes out of a continual dialogue on many different planes with other sexual persons. We encounter many images of what it means to be a human person and what it means to be masculine or feminine. And soon, hopefully rather than later, we find that in the human situation we must respect the different languages, the different life styles, and the different approaches people take to becoming human sexual persons. Along the way we make many decisions, some of them fortunate, others not so, but the dialogue continues throughout life. This is the challenge each one of us faces.

If technology does indeed create its own culture and gives it its orientation, then inherent in the new technology of human reproduction we have examined here should be the seeds of future human behavior and culture. Those trends, those seeds, were already apparent by the time we surveyed artificial insemination with all its ramifications. Having followed through with a survey of artificial wombs and test-tube fertilizations, superovulation and mercenary mothers, asexual reproduction and cloning as well as our ability to monitor and manipulate the unborn, we can now expand to some degree on the three basic trends we outlined at the close of our first chapter.

I.

It seems very evident from all we have seen that for man and woman both sexual intercourse and procreation are already, and will be more so in the years to come, distinct human activities. Human reproduction can now bypass sexual intercourse, and this is likely to become more common in the days to come. The decision to reproduce then becomes a fully free, human responsibility.

In expressing their deep dissatisfaction with the traditional churches' attitude toward human sexuality as a means to an end rather than as an end in itself, many people today have fallen into the trap of a subtle but fatal confusion. To maintain that reproduction is not the prime or sole justification for human sexuality, that procreation is only one, perhaps even accidental, element in the creative possibilities of the male-female dialogue, is not the same as subscribing to the "sex for fun" thesis. This has been very clearly emphasized by Joseph Blenkinsopp, a young biblical scholar. Alex Comfort, poet, novelist, and medical biologist, may regard human sexuality as the most important human sport, and others who accept the thesis of "sex as fun" do in a way view human sexuality as an end in itself. But when sex becomes a sport or fun and nothing else, it generally ceases to be fun. The playmate becomes a plaything; spontaneity, joy, and lightness are lost and the game becomes a sick contrivance. Despite this confusion, there is a sense in which we can and must affirm that sex is fun. Eros (or sexual expression in its fullest sense) and fun are closely allied, for both spring from a sense of joy, celebration, and spontaneity which is spoiled by any consciousness of purpose. Both, like the dance, are forms of contemplation.

But this carries with it certain risk. "Once sex and reproduction are separated," Dr. Robert S. Morrison of Cornell warns, "society will have to struggle with . . . defining the nature of interpersonal relationships which have no long-term social point . . . [and] seek new ways to ensure reasonable care for infants and children in an emotional atmosphere which lacks biological reinforcement. . . ."

Freed of its procreative function sexual intercourse takes on a fuller and more human meaning as a profound and rich means of interpersonal dialogue and communion. What does this hold for the normal maturing relationship of a young man and young woman today? From infancy on we have learned to live with sexual people, with young playmates and with our parents and others of the older generation. From them we have learned to engage in dialogue on a variety of planes and in many languages. In this world where sexual intercourse becomes a language of communion, young people are seeking ways of expressing themselves as sexual persons. More often than not they do not see this in terms of procreation or marriage or family. In some ways a comparison with an infant taking his first toddling steps or struggling with his first words seems appropriate for young people in high school or the first years of college. They are groping with a new language and are bound to make mistakes. Sometimes, when your year-old son totters on the brink of the stairs wondering in his budding mind whether to try crawling down them, the only thing a parent can do is look the other way. The incidence of premarital sexual intercourse and experience is rising to major proportions, and many young people are testing their sexuality by living together for weeks or even months without really seriously considering marriage. For older people sexuality and sexual intercourse have always been linked with marriage and procreation. But for a younger generation, and the gap may be one of only five or ten years, contraceptives and other developments have created new meaning in sexual intercourse. The premarital experiences of young people today are not seen in the plain context of hedonistic pleasure or mere genital sexual activity. It has gone beyond that narrow base, so that even in, and perhaps especially in, situations when young people live together for some time without thought of marriage, the prime concern is to work out one's own masculine or feminine personality in a profound concrete dialogue and experience.

Some "trial marriages" may mature into socially acceptable and formal marriages, but more come to an end with the two young persons finding that they are growing in different directions. Despite the complications of romantic love and the hope of a total giving of self to another which often confuse and agonize them, this experience may be necessary for a young man or woman to mature to the point where marriage becomes a reasonable life style. But just as learning to walk involves many a bruised head and skinned knee, so learning the language of the mature sexual person also entails many emotional traumas.

For most parents, sex education whether at home or at school means simply an introduction to the "facts of life." Even when this education tries to bring human sexuality into the context of human society and out of the biological and individual, it is always in terms of *past* social structures with absolutely no indication of any awareness of what is happening in human society as a result of our reproduction technology. The Pill may be discussed in a high school health course, but even the best teacher or parent shies away from facing the radical impact this development has already had on our relationships in and out of the family. The result is often a tremendous naïveté among young people, particularly girls. A growing number of young people are convinced that human sexuality arises through dialogue and that sexual intercourse may be justified in a deep personal relationship outside marriage. From this base, they frequently plunge into a relationship without the slightest thought that pregnancies do occur unless some precautions are taken. The idealism of youth is not always as practical as it could be. Even more serious in my mind is the dead end many young people fall into because they accept their parents' equation of masculinity or femininity with the experience of sexual intercourse. Because they have not been prepared for the new and deeper understanding of the male-female relationship, these young people equate the experience of sexual intercourse with the achievement of the full masculine or feminine ego. Having reached that minor plateau of experience, they often develop a wearied, bored blaséness about sexual intercourse which is devastating to the male ego particularly when found in a woman's reactions. Without realizing that human sexuality is not encapsulated in the genital many young women fail to develop their feminine personalities to include that coy, elusive allure that makes a real woman utterly feminine all the time and in every direction.

The revolutionary aspect of this is that we have in effect scrapped our whole classical image of man and woman and their individual role in society as well as their interrelationship to each other. Having demolished that image, we have yet to create satisfactory new images of man and woman, of what it means to be masculine or feminine in this new era. Such interesting experiments are emerging as the Danish "house-husband" where the male actually prefers to stay home and take care of the chores and children while his wife works at a very good, well-paid job she finds extremely fulfilling. This obviously is not for every man, by any means, but is this experiment, and countless variations on the theme, not helping in its own small way to evolve new images, and a pluralism of images for the two sexes within human society? Many husbands today find themselves helping with the housework or changing diapers, not because they have to but out of free choice. A growing number of husbands find themselves sharing what was once considered the wife's domain, even as wives are broadening their own experiences and modes of personal expression.

Marshall McLuhan has offered some engaging comments on this whole question of future images of male-female in his 1967 article for *Look* magazine. Among the most isolated and primitive tribes known today, those which still center on a free-roving hunter culture, digestive disorders, ulcers, and the usual civilized heart troubles are a rarity. McLuhan and Leonard feel that a major cause for this absence of our civilized diseases lies in the fact that these people make little distinction between the *ideal* qualities of male and female. Geoffrey Gorer, a noted British anthropologist, commenting on the Pygmies of Africa, the Arapesh of New Guinea, and the Lepchas of Sikkim, has noted that "Men and women [in these societies] have different primary sexual characteristics—a source of endless merriment as well as more concrete satisfactions—and some different skills and aptitudes. No child, however, grows up with the injunctions, 'All real men do . . .' or

'No proper woman does . . .' so that there is no confusion of sexual identities; no cases of sexual inversion have been reported among them. The model for the growing child is one of concrete performance and frank enjoyment, not of metaphysical, symbolic achievements or of ordeals to be surmounted. They do not have heroes or martyrs to emulate or cowards or traitors to despise; . . . a happy, hard-working and productive life is within the reach of all."

In the usual Western society, "being a man" involves aggression, and for McLuhan and Leonard this is, if nothing else, unhealthy. A recent Army study of the "nervous secretions" of combat soldiers in Vietnam has shown that an ordinary "peaceful" life in the United States can equal or exceed the dangers of enemy gunfire in terms of intestinal and nervous upsets. But current trends in human sexual patterns in the States contain a hopeful promise in the movement away from our present fragmentation of sexual roles.

Extremes usually create opposite extremes. And the highly specialized, aggressive, profit-conscious male of the industrial age has produced, at least as an ideal mystique and myth, the specialized woman. In an age which stresses the visual over all the other senses—an age dominated by technologies of the photographic and cineramic—the realities of femaleness have been abstracted from the context of everyday life and turned into something intensely "hot" in the McLuhan sense. Today, as we approach the demise of that visual era, we find grotesque and distorted extremes of the mythic female everywhere in the visual shock of "I Am Curious Yellow," "Without a Stitch," and the like, as well as in the utopian Amazons that grace the foldouts of *Playboy* and its imitators. Already the Playboy Bunny and Playmate are beginning to appear quaint and to stir a strong reaction among the American youth who see in these images only the death rattle of a past culture with its infantile fixation on breast and buttocks.

The sex symbols of the new age already poke fun at these exaggerations of our recent past: the boyish Twiggy and the effeminate hyperboles of Tiny Tim. McLuhan draws a comparison which equates the messages of Sophia Loren and Twiggy with those of a Rubens painting and an X ray. He finds in the X ray, not a realistic picture or specialized female, but rather a deep, involving image of a *human being*. From this vantage point and judgment, McLuhan and Leonard argue that both sexes are now moving toward a common humanity where the unalterable essentials of maleness and femaleness will hopefully assume their rightful importance and delight. Where "glamour," a form of armor, insulated and separated, McLuhan sees in the new male and female styles an invitation to dialogue.

While the cultural patterns of both sexes are unavoidably drawn into the maelstrom of change, as McLuhan has warned, "most men will have farther to go than most women in adjusting to the new life."

II.

No social institution is more basic in Western society than the monogamous or couple marriage. And no institution is likely to suffer more radical changes in the biological revolution we have detailed here than that familiar family structure of one husband, one wife, and several children.

The conjugal family, as we know it in the Western world, is a very late development. While ethologists like Konrad Lorenz and Desmond Morris argue in such books as *On Aggression, The Naked Ape,* and *The Human Zoo* that the pair-bonding observed among the monkeys, apes, chimpanzees, and humans is genetically established by mechanical selection as the sole basis for the conjugal human family and love, most anthropologists reject this as a gross oversimplification. There has never been one inherited pattern for marriage among men. There has, in fact, been quite a variety of marital patterns for stabilizing the relationships of male and female, particularly in terms of procreation. Each of these patterns was invented in the course of human affairs as a specific adaptation to meet the demands of a special environment. Considering the totally new environment created by the new embryology, it is most likely that we will evolve new patterns of human relations more amenable to the exigencies of our new technologies of reproduction and their repercussions for the individual sexual person.

I find it very difficult, if not impossible, to accept claims such as those set forth by Eleanor Garst in *The Futurist* and *Marriage* magazine that these new developments will soon render the monogamous couple marriage passé and a rare relic of the past.

Human institutions are simply not that fragile, and the monogamous couple marriage, despite its many fluctuations and partial failures, is well established in the human psyche of Western man. I would, however, agree that these developments are going to *modify, and very likely modify drastically,* our experience of the couple marriage. Their most likely area of impact appears to be already evident in our changing concept of the exclusivity of the couple. Modifications are already appearing in this aspect through the appearance and growing social acceptance of alternate forms of male/female relationships and parenthood alongside the couple marriage, which itself occasionally and consciously incorporates certain modifications in the traditional concept of fidelity.

Margaret Mead, a world-renowned and respected figure for half a century in anthropology, has advocated a major modification in our society, the "trial" or "provisional" marriage. Considering the complexities of modern life and the fact that few young people, even in the early or mid-twenties are really prepared for a lifetime marriage, Dr. Mead has suggested that society recognize an agreement between two young people to live together for a year or two, provided they use an appropriate contraceptive and have no children. This period would allow the young man and woman time to become really acquainted with each other, to share more deeply and learn whether they really want to share their lives on a more permanent basis by starting a family. Professor Mead argues that this period of adjustment would allow the evolution of a much more stable foundation for a lasting marriage should the couple decide to enter into a "parental marriage." In effect, informal trial marriages are already the case for a growing number of young people. But again it is worth emphasizing that this is not a matter of testing one's sexual compatibility; it goes far beyond that rudimentary concern into the in-depth mutual development of two sexual personalities.

A quite different type of "trial marriage" has been urged by the director of the Marriage Counseling Service of Wayne, Michigan, Circuit Court. In this proposal for reducing the number of hasty, precipitous marriages, Edward J. Staniec includes in his sixty-day "trial marriage" all but one ingredient of a bonafide marriage, sex relations, which actually make up only a small part of any marriage. For the first month, the engaged couple would remain completely apart from each other. They would have no contact, no telephone calls, no exchange of letters—nothing but memories to sustain and test their love. At the end of each week, they would fill out a self-analysis questionnaire, asking themselves how much they really missed their beloved, whether or not they had been secretly looking around and enjoying mental flirtations with others, whether they really trusted the love of their beloved. Having passed successfully this first part of the trial, the couple could then spend a second month together, testing the "boredom factor." During this period, the couple would spend all their time alone together, except for work and sleep. As soon as he finished work, the young man would rush home to share a dinner cooked by his fiancée. Evenings and weekends would also be spent alone together at home, watching television, reading, or just talking and sharing. No physical contact, not even a good-night kiss would be permitted during this month. Nor would the couple be allowed any company or evenings out on the town. The key question to be answered by this trial is how compatible are the couple likely to prove when faced with the humdrum of everyday married life. If the engaged couple honestly and successfully evaluated their relationship, they could then enter into a full marital life. Though this proposal is not likely to attract many altar-bound couples, it is a creative proposal worth serious consideration.

The human family began in man's early tribal culture as an "extended family," embracing a large number of people, aunts, uncles, cousins, father and mother, grandparents, and for each woman perhaps a dozen or more children. Polygamy, polyandry, and countless variations were tried at one time or other. Today the human family has evolved down to a bedrock, one father, one mother, and two, perhaps three children. The "nuclear" form of family life has been criticized by many including Margaret Mead as "one of the worst forms [of family life] ever invented." In the past two or three decades, when it has come to dominate American life, the nuclear family has produced intense loneliness and alienation. Much of this loneliness stems from a reduction in the supporting relationships which characterized the extended family even in its more sophisticated and modified versions of the first quarter of this century. But perhaps equally important as a negative factor has been the reduction in the number of children per couple. With the growing awareness of the population pressure, this limitation of procreation is likely to increase, perhaps even to the point where many couples will consider marriage but without children.

In an address to a 1969 conference sponsored by the American Foundation of Religion and Society, Margaret Mead saw definite hope for the family in the new technology. She argued that the population explosion, combined with new methods of birth control, is creating a situation where society should eliminate its pressures for all people to marry and have children.

> At present we're ferociously over-married. We think everybody should be married all the time, from puberty till death. The idea has even pervaded professions formerly devoted to celibacy.

Professor Mead foresees a solution in the evolution of a new pattern for marriage, the "cluster family." This would not be a reversion to the extended family of man's tribal days, as McLuhan wrongly argues (though I do accept his proposition that mankind is entering a type of culture which can best be described as a global tribe because of the instantaneousness of electronic communications). Mead's cluster family would include single adults and elderly people as well as young couples with children. As in the tribal extended family all the adults in the cluster would have some role in rearing the children. (In some respects the Israeli experiment with the kibbutzim anticipates the cluster family.) The cluster family, however, is not in Mead's way of thinking a group or communal marriage, for despite the widening of relationships within the cluster closed monogamous units would continue within the broad group.

But young people today who wish to explore and develop new forms of family and male-female relationships must realize that this venture means not merely a revolt against the traditional culture of Western society, but also a painful overcoming and maturing of the subtle emotional and psychological conditioning they have received from birth. Robert H. Rimmer has explored some of the tensions and complications that arise when young people enter this venture. In his 1967 novel, *The Harrad Experi-*

ment, Rimmer records in journal form the groping development of six students attending an experimental college in Massachusetts. The conflicts, tensions, and painful maturations experienced by Beth, Harry, Sheila, Stanley, Valerie, and Jack, coed roommates at Harrad College, are very instructive. Similar tensions and emotional conflicts appear in slightly different form in the sequels to Harrad, *The Rebellion of Yale Marratt* which explores the possibilities of a comarital or bigamous relationship between three consenting adults and *Proposition 31* which traces the birth of a group marriage. Other novelists have found in this cultural evolution fertile ground for creative and serious documenting of the human situation. The movie *Bob and Carol & Ted and Alice,* Aldous Huxley's *Island,* B. F. Skinner's *Walden Two,* and Austin Tappen Wright's *Islandia,* among others, have taken up the theme with some interesting insights. The utopian, and somewhat unrealistic, character of much of this writing, however, becomes apparent in the fixation of Rimmer and others on the total mutuality of group marriages. In *The Harrad Experiment* and *Proposition 31,* the males balance out the females perfectly and their mutual agreement is eventually worked out. In actual life, it may not be that simple and the realism of the "co-marital" relationship may be more commonly along the less idealistic lines of Yale Marratt.

Contrary to the views of some sociologists who see in our present culture the death rattle of the family, McLuhan believes that the family may actually be moving into a Golden Age. With so much room available for experimentation and flexible explorations, marriage, he believes, will occur much later in life than ever before. Family units in the future may organize loosely into "tribes" without fragmenting into little encapsulated nuclear worlds. As things stand today, the nuclear family often has nowhere to turn for advice and encouragement when in need save to professional counselors. The informal tribe of the future will, in McLuhan's mind, provide the necessary support and love of a sounding board far more effectively than our present professionals. "Marriage—firmly and willfully welded, centered on creative parenthood—may become the future's most stable institution," according to McLuhan and Leonard. The "togetherness" of recent decades has been based on stereotyped conceptions of each element in the family. The new family, the family that is already evolving, is integral and deeply involving in a wholly new sense. It may provide the ideal unit for personal discovery, and it will provide this germinal ground if within the family we recognize the seemingly infinite possibilities of being human. Each child then would be viewed not as some inescapable threat to our stereotype images of child, male, female, father, mother, but rather as the trigger to making us aware of change, as an encouragement for all the members of the family to enter fully into *the creativity and stability of constant change*. Marriage and familial roles then would finally be freed from the compulsions and restrictions surrounding "high-intensity *sex*." The result would be a more sensual and integral approach to marriage.

In groping for new forms of family life more suited to our technological culture, man may also find the celibate marriage a valuable addition to his emerging cultural pluralism of male-female relationships. In the antisexual mentality of past generations, marriage was justified only as a means of increasing the number who could share in the Kingdom of God. Procreation justified marriage and the use of sexual intercourse; it gave marriage its foundation. Today, with sexual intercourse almost totally separated from procreation, marriage is finding a new basis in the lasting interpersonal relationship of creative men and women. Many would maintain, and rightly so, that for most humans their development as sexual persons is very much linked with their experiences with other sexual persons in the dialogue of sexual intercourse. However, this may not and cannot be a universal requirement. In exploring a *New Dynamics in Sexual Love: A Revolutionary Approach to Marriage and Celibacy,* philosopher Robert Joyce and his wife Mary outline what they call the "celebrational" or celibate marriage, in which sexual intercourse is a very minor and seldom if ever exercised element in the interpersonal creative relationship of a man and woman. Basically, I find the Joyces' proposal too reminiscent of the "brother-sister" marriage recommended in the Middle Ages in a blatantly antisexual climate and of Augustine's (and Teilhard de Chardin's) view that we are headed toward an asexual, or at least celibate life hereafter. Nevertheless, this approach may find some, even considerable acceptance. I would suspect that few married couples are up to its mature demands. But there is no denying that a married couple might well mutually evolve in this direction.

Communal marriages are not a new phenomenon on the American or European scene, although recent popular magazines imply that this is something totally new in human experience. Some communal marriages are based on a strictly monogamous couple structure in which several married couples set out to live a community life, very reminiscent of the extended families of the past generations which included aunts and uncles and other relatives all living under one roof. In rejecting the standards of an affluent society where even out-of-the-way luxuries are taken for granted, most communal groups are seeking a new appreciation of material things by letting loose what their parents have fought so hard for to spare them the trials and tribulations of this life. Some of these groups are more creative, exploring the possibilities of a complete communal life somewhat in the direction of an evolved combination of polygamy and polyandry.

Until very recently, and generally speaking even today, fidelity in marriage has been considered quite clear cut: for most people it means a total complete sexual (genital) faithfulness which excludes sexual intercourse, and even its serious preliminaries, with anyone except your spouse. But if we no longer base marriage on the procreative function and sexual intercourse is a valuable and fruitful mode of interpersonal communion, dialogue, and growth, then can we continue to base our concept of fidelity in marriage on sexual exclusivity? Naturally, there are many different reactions to this question, because there are many different emotional and psychological conditions to which a person can be exposed during his formative years. The American scene, like the world in general, will always remain an ever changing pluralistic pattern of ethical judgments, social practices, and variations for the individual conscience.

Human sexuality as we have known it for centuries may soon be dead according to Marshall McLuhan. Already our sexual conceptions, ideals, and practices are being altered, often beyond recognition. Marriage and the family are both shifting into new dimensions and worlds. There will likely be many surprises in the days to come and the meaning of being a boy or a girl, a man or woman, husband or wife is likely to be quite different from what we have experienced thus far in human history.

Nowhere is this truer than in the fading romanticism of modern youth. While one negative aspect of this may well be the loss of a very precious sense of wonder, awe, and joy in the sexual relationship of boy and girl, there is little doubt that the romantic myth is dying.

As a way of selecting a spouse, romance ("In all the world, you are the only girl for me") never worked very well. Back in the 18th century, Boswell may have felt some shock at Dr. Johnson's answer to his question: "Pray, Sir, do you not suppose there are fifty women in the world, with any one of whom a man may be as happy, as with any one woman in particular?" Johnson replied: "Aye, fifty thousand." The future may well agree with Dr. Johnson. It is difficult to play the coy maiden on a daily diet of contraceptive pills. And the appeal of computer dating suggests that young people are seeking out a wide and quite practical range of qualities in their mates—not just romance or high-intensity sex appeal. Here, in fact, may be the electronic counterpart of arranged marriage.

As Romantic Love fades, so may sexual privacy. Already young people shock their elders by casually conversing on matters previously considered top secret. And the hippies, those brash pioneers of new life patterns, have reverted—boys and girls together, along with a few little children—to the communal living of the Middle Ages or the primitive tribe. It is not uncommon to find a goodly mixture of them sleeping in one room. Readers who envisage wild orgies just don't get the picture. Most of the hippies are *not* hung up on sex. To them, sex is merely one of many sensory experiences. It is available when desired—therefore perhaps not so desperately pursued.

In some ways human sexuality is entering for the first time the adult world, and here I would disagree with McLuhan, who concludes the above three paragraphs with the judgment that sex is "returning to the adult world." The distinction is vital, I feel, for we are in no way returning full cycle to an earlier stage of human development. We are engaged in the evolution of mankind and people, not in a cyclic fixation with the long awaited return to a Paradise Lost. Genital sex is indeed becoming less central to the young. Its focal intensity is dissipating even as it becomes more accessible. I recall the experience of a contemporary of mine whose eighteen-year-old daughter informed him she would be spending the summer touring Europe on a bicycle. "Alone?" "No, Dad, I'll be O.K. Joe and I are going together." Raised eyebrows but fortunately no verbal explosion. But even this was noticed by the daughter who chided her father: "There you go, Dad, assuming Joe and I are going to be spending all our time in bed. We're going to see Europe, the countryside, the museums, the people, the customs . . . not sleep together. We're only friends!"

With marriage coming, hopefully, later in life, it will become a much more serious and mature development. McLuhan suggests, just as serious as divorce, proposing that some couples may even wish to write up a legally binding separation agreement (to be revised now and then as financial and parental situations change) as a precondition to marriage. Anticipating the possible unpleasantness of a divorce even before marriage McLuhan feels may make divorce far less likely.

A number of points raised by McLuhan, including his gross oversimplification of the death of romance in human relations, can be argued, but these paragraphs likewise pose some serious points for meditation when we consider the future of the family.

In a chapter entitled "Looking toward the 21st century," Vance Packard, the author of *The Sexual Wilderness*, admits that "the institution of marriage is obviously in need of modifications to fit the moderns needs." In this context he discusses 7 possible patterns of marriage or "near-marriage:"

1. Serial mating, also known as serial monogamy and serial or consecutive polygamy, perhaps with marriage contracts renewable every five years;

2. Unstructured cohabitation, stable open non-marital relationships which were once limited very much to the lower classes but are now fairly common on the college campuses and among the upper classes;

3. Mutual polygamy with its more informal variant, "flexible monogamy" as urged by Phyllis and Eberhard Kronhausen where friendships can include sexual experiences within the context of stable monogamous marriages;

4. Single-parent marriages by intent, or bachelor parents (with cloned offspring);

5. Specialists in parenthood as urged by Margaret Mead and others who feel that in the near future parenthood will be limited to a small percentage of families whose prime function will be child-bearing and rearing;

6. Communal living involving several adults of both sexes in a stable relationship;

7. Legalized polygamy for senior citizens.

Vance Packard concludes this list by suggesting that what will more likely occur in the near future will be the acceptance by society in general of *an open pluralism involving most if not all of the above 7 options.* I would concur with this appraisal, but add to the list at least two other alternatives: the "celibate or celebrational marriage" proposed by the Joyces and variations on the polyandrous theme should we upset our balances of the sexes with sex determination so that we end up with 150 males for every 100 marriageable females as the Eskimos have as a result of their selective infanticide. It is also likely that an individual will experience more than one type of relationship among these possibilities and that these experiences will be formalized in degrees varying from simple mutual private consent to public announcements and formal weddings.

III.

Finally, we come to the question of an evolution in human values as mankind meets the challenges of utopian motherhood.

In a recent speech, Dr. William H. Stewart, U. S. Public Health Surgeon General, warned us that "today we are on the threshold of realizing what no one had dared to dream a few short decades ago—effective human intervention to prevent or reverse the tragic results of events occurring before birth." As our scientific knowledge grows, and our technological skills increase by leaps and bounds, each of us will have to face the ethical questions posed by our ability to invade and manipulate the heretofore inviolable domain of the uteronaut. But, as Dr. Frank Ayd has pointed out in the *Medical-Moral Newsletter*, "neither the [1964] Declaration of Helsinki [drafted by the World Medical Association] nor any other code of ethics governing human experimentation includes guidelines for experimentation of germ-plasm or prenatal human life." The official churches have maintained a rather strange silence on these issues, either ignoring their reality or preferring to be distracted by such nonproblems as the contraceptive pill. Individuals in the street are thus reduced to appraising these techniques and developments from their own individual perspectives, without the aid of guidelines from those on whom they were accustomed in the past to depend for a distillation of age-old wisdom and experience.

The fundamental consideration in the moral or religious life of the Judaeo-Christian tradition has always been the fullest possible growth of the individual as a mature loving person. But this growth must always take place in dialogue with the other, with society, and hence a mutual responsibility emerges in any moral judgment we may make.

We have mentioned the possibility of premarital sexual intercourse. The term is inappropriate for the present situation, but we have no other to replace it. Inappropriate, because frequently this experience is not entered into as a prelude to marriage, and marriage may even be consciously and mutually excluded by the parties. Returning to our example of the infant learning to walk and run, and considering the "hothouse environment" practically every high school student and most college students find themselves in, it seems apparent purely from a psychological viewpoint and without any recourse to religious codes of ethics, that many young people are not ready to handle the full language of the bodied-human. Throughout their adolescence and teen years, both young men and young women are in the process of developing and maturing as persons. They are *not yet sure enough of themselves as individuals, as masculine or feminine persons,* to engage in the fullest expression of sexual love. Ashley Montagu summed this up succinctly in his latest work, *Sex, Man, and Society:*

> With the pill, premarital sex without any of the anxieties usually associated with it or with the birth of children becomes for the first time possible, and hence the principal barrier against it is removed. But with the removal of this barrier the responsibilities involved in

this particular relationship are maximized beyond anything that has hitherto been anticipated or required. For once the barrier has been lowered, the danger of the debasement of this delicate, this tender, this most sensitive of all human relationships is greatly increased. Hence, no one should ever think of entering into such a relationship who is incapable of behaving responsibly in it. Responsibility to others is something one must learn. It is not something one is born with. It is here that the schools must assume *their* responsibility, for it is in the schools that the parents of future generations must be prepared in the meaning of sex and responsibility.

The criteria then become a mutual and completely frank honesty, a mutual concern for personal growth and maturation, plus a mutual recognition of the possible outcomes and consequences of one's decisions.

In the Middle Ages people were greatly disturbed by the Church's insistence that, contrary to century-old custom, the bride would now have to give her consent freely for a marriage to be valid. Many parents thought it absolutely immoral to suggest that the bride give her consent. Today we would think it absolutely immoral were the bride not to give her consent. In a pastoral society where great prize was laid on the dowry of a virgin daughter and the legitimacy of heirs made fidelity a prime concern, premarital sexual experience was very limited. How do we view it today? Personalism and individual responsibility are rapidly replacing the old morality imposed on us by society, our parents, and other outside sources. True, we are still conditioned to some extent by our families and our place in society, but more and more the individual must make decisions for himself, thoughtfully considering and pondering his own growth as a human person, his creative contribution and responsibility to others, and his relationship with a transcendent God.

On the question of marital fidelity and its possible evolving meaning, it is worth noting that individual people vary quite significantly in the psychological and emotional reaction on this point, primarily because of their cultural conditioning. A woman who has been subconsciously conditioned to think of the stability of her marriage in terms of complete sexual fidelity will find it difficult to accept a co-marital situation, or a passing sexual encounter on the part of her husband with another woman. We might recall here some remarks made by Della and Rustum Roy, husband and wife members of the Sycamore Community in Pennsylvania and the authors of *Honest Sex: A revolutionary sex ethic by and for concerned Christians:*

> Further, it is our considered judgment that just as the twenty years following World War II saw a revolution in the attitude to premarital sexual activity, the problems of the next twenty years will be concerned very largely with post-marital sexual behavior. Just as there has been a phenomenal increase in the total premarital sexual intimacy of all kinds, we foresee that the moral and ethical dilemmas of Americans for the next decades will probably center largely on developing creative and workable patterns of co-marital activity (including but not limited to coitus). We will use this term *co-marital* (first suggested to us by the Reverend William Genne of the National Council of Churches) to describe without the pejorative connotation of the term *extramarital,* any man-woman relationship, and/or sexual expression thereof, which exists alongside of and in addition to a marriage relationship. Such relationships are basically not competitive with the marital relationship; they may have a neutral or even a positive effect on it. The term *extramarital* will still be used to describe situations which are not as clearly noncompetitive.
>
> This problem is not one which affects only the avant-gardes; nor is it confined to the least responsible sections of society. It is sometimes a problem among the most responsible self-aware groups including communities of Christians where persons are intimately bound to each other by the deepest loyalties. Indeed, it is not without significance that the Oneida community—one of the most prominent American experiments in a communal way of life—gave much thought to sex ethics and even to the appropriate sexual expression for such relationships. A communal life for this group logically included communal sexual relations with all members of the community on a strictly equal basis.
>
> The question should perhaps be asked in a fresh light: What rules guide the sexual intimacy between one person and another not his spouse, and to what situations does the application of such rules lead?

Few married couples today and certainly fewer newlyweds, I suspect, would have the emotional and psychological maturity and stability to accept a fully co-marital situation. But what of the less permanent co-marital relationship which in the past has been formally and commonly condemned as immoral but today, in the context of interpersonal dialogue, takes on new overtones? The Roys argue that

> It is utterly ridiculous to say on the one hand, "Greater love hath no man than this, that he lay down his life for his friends," and to assert immediately that it is impossible and unnatural for a man (or woman) to agree to share his (or her) spouse with another. We are claiming then that no black-and-white case can be made against sexual intimacies (including coitus) between persons not married to each other.

Not a few marriage manuals and books treating of marriage and dating would agree with this proposition and carry it even further. In a more moderate vein, however, is the approach taken by a British classic edited by Alastair Heron, *Towards a Quaker View of Sex,* published in 1963:

> This is probably the point at which to mention the so-called "triangular" situation. This is too often thought of as a wholly destructive and irresponsible relationship, the third party being at the very least an intruder and at worst an unscrupulous thief. Its portrayal thus in fiction and drama no doubt contributes to the stereotype. . . . Not sufficient recognition is given to the fact that a triangular situation can and often does arise in which all three persons behave responsibly, are deeply conscious of the difficulties and equally anxious to avoid injury to the others. . . . It is worth noting that in the two-woman: one-man situation, the very happiness of the marriage may attract a young girl or a sensitive and responsible woman. . . . By the same token, it could surely help a nervous youngster to fall in love with a happily married man.

Perhaps one development in the coming years will be a more open acceptance of something which our society's Victorian blindness prefers to ignore.

This is *not* to endorse promiscuity. It is rapidly becoming evident that while there is a much more open attitude among young people to sexuality and certainly more premarital and "extramarital" sexual expression on the part of younger people, this is in a much more personal context of dialogue and profound relationships. Promiscuity, bed-hopping and spouse-swapping are nothing but hedonistic self-seeking; they are destructive of the human person because they are egotistic, immature, adolescent and very often entered into without real love. As Joseph Fletcher, the controversial author of *Situation Ethics,* put it rather briefly:

> Adultery, for instance, is ordinarily wrong, not in itself but because the emotional, legal, and spiritual entailments are such that the overall effects are evil and hurtful rather than helpful—at least in our present-day Western society. But there is always the outside case, the unusual situation, what Karl Barth calls the "ultima ratio," in which adultery could be the right and good thing.

The past fifty years have undoubtedly seen more radical changes in man's understanding of his own sexuality and its ethical appraisals of it than the preceding three thousand years. And the situation gives evidence of becoming even more revolutionary in the decades ahead.

One prime development is already evident: sex is becoming less genital and merging more with the rest of life. Human sexuality must take its place by diffusing throughout the whole spectrum of human experience. As McLuhan has stressed so often:

> The coming age, linked by all-involving, instantaneous, responsive, electronic communication, may seem more "tribal" than "industrial." The whole business of sex may become again, as in the tribal state, play—freer, *but less important.*

We should also consider other areas of human values as they are likely to be affected by the new embryology, particularly in terms of the child and his place in society. Is this necessarily linked with the couple marriage? Today a single person can adopt a child; tomorrow a young single woman might ask her doctor to help her conceive by artificial insemination, or a young man might decide to clone a couple of sons. One-parent families—are they so inconceivable in terms of our blending and mingling of the traditional male-female roles? A single mother might have several brothers and male friends so that her child would be exposed to several different males from whom he could develop his own image of man. Society is already taking tentative steps that would facilitate such a development. Our latest welfare proposals are paying special attention to expanding day-care centers so that mothers in fatherless homes can work. The day may soon come when the isolated nursery school will be replaced by an American-style kibbutz in every suburban development and every urban housing complex. Again the question is one of accepting a pluralism in life styles.

Sooner or later we will have to face the question of a limitation on the population and the individual's right to reproduce. Whether we will come to an outright eugenic program is one question, but the possibility of legal restrictions on reproduction in general are well within the foreseeable future, perhaps only decades off. This situation will force us to ask how inalienable is the individual's right to reproduce? Does the good of society prevail in certain cases over the good of the individual? And can such a principle be applied here?

Will communal marriages become more common, and more accepted? What might be their impact on our image of father and mother? How is the communal marriage likely to affect our concept of fidelity in marriage? Are we likely to reach the stage observed by Margaret Mead in some South Pacific cultures where all women of the village are equally mother to every child and all the children of a particular age group are brother and sister to each other? In a communal marriage should we recognize any distinction among the parents, perhaps by indicating: "Sired by . . . ; other fathers . . . ; Born of . . . ; other mothers . . . ?"

Finally, how will we come to view the institution of marriage itself *in the context of process?* In the past the validity and reality of a marriage was simple to judge. One required only a couple of simple answers: did the bride and groom freely consent before a religious or civil official, and was the marriage consummated? If so, the couple was married for life. But within the process context, we find an encounter and dialogue between a man and woman developing through many stages. It matures slowly and not infrequently the couple grow apart and their relationship is not affirmed in a marriage ceremony. But if the process can abort before a public marriage, might it not also abort afterwards? If so, does this mean that a marriage has terminated. In actual fact, such a marriage is not a full marriage, it is only in the process of becoming, of being born as a full human relationship, and it happened to abort along the way. Only at death is it possible to say of a man and woman they were truly married. It seems within the realm of possibility, and there are already some distinct indications even in the Roman Catholic Church, that many Christians who have resisted any acceptance of divorce may eventually recognize that, regrettable as it is to accept the fact that a marriage in process has not and will not reach its goal, accepting that fact and freeing the couple to seek their maturity in other relationships may be more Christian than trying to deny reality.

Throughout this discussion of the new embryology, it should be evident that society faces a major shift toward pluralism, without the stigma that variations have often brought in the past. Slowly, agonizingly, mankind is coming to recognize, even in the area of interpersonal relations and procreation, the truth of that old adage that one man's meat may be another man's poison.

To deal with the human phenomenon is to grapple with the totally unpredictable. Some of the techniques and trends we have outlined here will undoubtedly abort and never become more than a peripheral groping that proved unsuccessful. Undoubtedly some of the developments sketched above should be ruled out by mankind. But others will flourish and expand to become major forces in the coming generations. Still other techniques are likely to take unexpected turns and twists so that in the years to come we may hardly recognize them.

Throughout all this fluid indecision three facts remain indisputable. First, the ruinous dualism of body versus soul in man, which has resulted in a dichotomy between our sexuality and the rest of our lives, will continue to evaporate until nothing remains of the mentality which only fifty years ago could view sexual intercourse, even in marriage, as a "thing filthy in itself." Second, the major movement toward an emotional, psychological, and social acceptance of a pluralism without recrimination in life and marriage styles will continue to accelerate and spread. Finally, human sexuality will at long last be humanly integrated into our lives for what it is in reality, a shimmering aura coextensive with the emerging human personality engrossed in the lifelong task of coming to birth as a mature sexual person. In that dynamic context the dialogue of sexual intercourse and procreation will find lasting, if constantly changing niches.

A SELECTED
BIBLIOGRAPHY
WITH SOME
ANNOTATIONS

(Editor's note: I have avoided the use of extensive footnotes and documentation in the text in order not to burden the average reader. However, I realize that some readers will want to know the source of some particular fact or perhaps explore further the ideas expressed by some of the scientists and scholars I have quoted. To facilitate this exploration I have annotated the following bibliographic entries to indicate their general content and relevance.)

ANDERSON, KENNETH N. "Can Man Be Modified to Live in Space?" *Today's Health*, November 1963.

ANDREWS, JAMES F. "Jesuit Denies Abortion in Early Pregnancies." The *National Catholic Reporter*, November 27, 1968.

———. "Theologians Add Little to Abortion Debate." The *National Catholic Reporter*, December 4, 1968. These two articles report in detail on the first International Conference on Abortion with particular attention to the revolutionary arguments of Jesuit Joseph Donceel.

ASIMOV, ISAAC. "Pills to Help Us Remember?" The *New York Times Magazine*, October 9, 1966.

AUGENSTEIN, LEROY. *Come, Let Us Play God.* New York: Harper & Row, 1969. A popular, readable exposition of genetic engineering and death control in terms of their human implications.

AUSTIN, C. R. *The Mammalian Egg.* Oxford: Blackwell, 1961.

——. "The 'Capacitation' of the Mammalian Sperm." *Nature.* 170:326, 1952.

AVERY, T. L., C. L. COLE, et al. "Investigations Associated with the Transplantation of Bovine Ova." A four-part survey in the *Journal of Reproductive Fertility*, 1962.

AYD, FRANK J., JR. "'Wanted' Children and the Population Crisis." The *Medical-Moral Newsletter*, May 1968. Summarizes Dr. William B. Shockley's proposal for eugenics and comments on this.

——. "Experimentation on Prenatal Human Life: Ethical Considerations." The *Medical-Moral Newsletter*, June 1969. Four-page survey of how both medical and religious groups have avoided proposing any ethical guidelines for prenatal experimentation.

BALFOUR-LYNN, STANLEY. "Parthenogenesis in Human Beings." The *Lancet*, June 30, 1956.

BELL, ROBERT R. "Some Emerging Sexual Expectations Among Women." A paper delivered at the June 1967 meeting of the American Medical Association.

BENJAMIN, HARRY. *The Transsexual Phenomenon.* New York: Julian Press, 1966. The only thorough and medically sound review of this current problem presently available.

BERNARD, JESSIE. *The Sex Game.* Englewood Cliffs, N.J.: Prentice Hall, 1968. A study of the many forms and levels of intersexual communication.

BERNHARD, ROBERT. "Strange rat 'race' leads to new biology theory." *Scientific Research*, March 17, 1969. Summary of Dr. Mintz's work with embryonic combination of different rat strains to produce allophenic offspring.

BERCHER, RUTH and EDWARD. "Every Sixth Teen-age Girl in Connecticut—." The *New York Times Magazine*, May 29, 1966.

BERRILL, N. J. *Sex and the Nature of Things.* New York: Dodd, Mead, 1953. A popular history of animal sexuality by a leading biologist.

BHATTACHARYA, B. C. "Pre-arranging the sex of offspring." *New Scientist*, October 15, 1964.

BLENKINSOPP, JOSEPH. *Sexuality and the Christian Tradition.* Dayton, Ohio: Pflaum, 1969. An excellent study of sexuality in the biblical tradition and its place in Christian culture.

BOROWITZ, EUGENE B. *Choosing a Sex Ethics.* New York: Schocken,

1970. A personally slanted but very informative survey of the evolution of Jewish morality over the centuries.

BRADBURY, RAY. "A Serious Search for Weird Worlds," *Life*, October 24, 1960.

BRENTON, MYRON. *The American Male*. New York: Coward McCann, 1966.

BRINSTER, R. L. "Developing Zygote." In: *Reproductive Biology*, edited by Howard Balin and Stanley R. Glasser. New York: Excerpta Medica, 1970.

BRONSTED, H. V. "The Warning and Promise of Experimental Embryology." *Bulletin of the Atomic Scientists*, March 1956.

BUNGE, R. G. "Further Observations on Freezing Human Spermatozoa." *Journal of Urology*, February 1960.

CAVAN, RUTH SHONLE, ed. *Marriage and Family in the Modern World*. New York: Crowell, 1960.

CHANG, M. C. "Development of parthenogenetic rabbit blastocysts induced by low temperature storage of unfertilized ova." *Journal of Experimental Biology*, 125:127–149, 1954.

——. "*In Vitro* Fertilization of Mammalian Eggs." *Journal of Animal Science*. 27, Supplement 1:15–22, 1968.

COMFORT, ALEX. *The Nature of Human Nature*. New York: Harper & Row, 1966. Excellent study of animal and human sexual behavior, human variability, and our prospects for the future.

CROW, JAMES F. "Mechanisms and Trends in Human Evolution." *Daedalus*, Summer 1961.

DAVIDSON, R. M. "Taking Charge of Evolution: What Next in Biology?" *Current*. March 1969.

DELGADO, JOSÉ M. R., *Physical Control of the Mind*. New York: Harper & Row, 1970. A thorough summary of Dr. Delgado's pioneering experiments with electrical stimulation of the brain.

DE MARTINO, MANFRED, ed. *Sexual Behavior and Personality Characteristics*. New York: Grove Press, 1966.

DOBZHANSKY, THEODOSIUS. *Mankind Evolving*. New Haven: Yale University Press, 1962.

——. *Heredity and the Nature of Man*. New York: Harcourt, Brace and World, 1964.

DUBOS, RENÉ. *Man Adapting*. New Haven: Yale University Press, 1965.

——. "Humanistic Biology." *American Scientist*, 53, 1965.

——. *So Human an Animal*. New York: Scribners, 1969.

——. *The Dreams of Reason: Science and Utopias*. New York: Columbia University Press, 1961.

DURHAM, M. "Experiments in Marriage: Group Marriages in Sweden and Denmark." *Life.* August 15, 1969.

DZIUK, P. J. "Egg Transfer in Cattle, Sheep, and Pigs." In: *The Mammalian Oviduct.* Edited by E. S. E. Hafez and R. J. Blandau. The University of Chicago Press, 1968. An excellent review.

EDWARDES, ALLEN and R. E. L. MASTERS. *The Cradle of Erotica.* New York: Julian Press and Lancer Books, 1962.

EDWARDS, R. G., B. D. BAVISTER, and P. C. STEPTOE. "Early Stages of Fertilization in Vitro of Human Oocytes Matured in Vitro." *Nature.* 221:632, 1969.

EDWARDS, R. G., *et al.* "Preliminary Attempts to Fertilize Human Oocytes Matured in Vitro." *American Journal of Obstetrics and Gynecology,* 96:192, 1966.

EDWARDS, ROBERT and RICHARD GARDNER. "Choosing Sex Before Birth." *New Scientist,* May 2, 1968.

EHRMANN, WINSTON. *Premarital Dating Behavior.* New York: Holt, 1959.

EISENBERG, LUCY. "Genetics and the Survival of the Unfit." *Harper's,* February 1966.

ENOS, J. PERRY, ed. *The Artificial Insemination of Farm Animals.* New Brunswick, N.J.: Rutgers University Press, 1952.

FERKISS, VICTOR C. *Technological Man: The Myth and the Reality.* New York: G. Braziller, 1969. A major analytical work detailing the impact of technology on human life and culture.

FISHER, ALAN E. "Chemical Stimulation of the Brain." *Scientific American.* June 1964. Report on experiments with injection of sex hormones and other substances into localized regions of rat brains.

FLETCHER, JOSEPH. *Situation Ethics: The New Morality.* Philadelphia: Westminster Press, 1966.

FRANCOEUR, ROBERT T. *Evolving World; Converging Man.* New York: Holt, Rinehart and Winston, 1970. Contains a detailed explanation of the author's latest attempt to draw out of various scientific and theological insights a new process image of human nature and of man. Very helpful in appreciating the philosophical and theological implications of the research on human sexuality and personality discussed in this present book.

GAGNON, JOHN H. "Sexuality and Sexual Learning in the Child." *Psychiatry,* August 1965.

GARRIGAN, OWEN. *Man's Intervention in Nature.* New York: Hawthorne, 1966. Somewhat dated but still useful thoughts on eugenics and mind manipulation by a biochemist-priest.

GERARD, RALPH W. "What Is Memory?" *Scientific American*, September 1953.

GREGOIRE, A. T. and R. C. MAYER. "The Impregnators." *Fertility and Sterility*, 16:1, 1965. The story of William Pancoast's experiment with human artificial insemination.

GURDON, J. B. "Transplanted nuclei and cell differentiation." *Scientific American*, November 1968. Cloning of frog embryos.

GUTTMACHER, ALAN F. "The Role of Artificial Insemination in the Treatment of Sterility." *Obstetrical and Gynecological Survey*, December 1960.

HAFEZ, E. S. E., ed. *Reproduction in Farm Animals*. 2nd edition. Philadelphia: Lea and Febiger, 1968.

——. "Animal Reproduction and Artificial Insemination." *Science*, November 8, 1968.

——. "Fertility and Sterility." *Science*, November 29, 1968.

——, and R. J. BLANDAU, ed. *The Mammalian Oviduct*. University of Chicago Press, 1968. Very useful surveys of embryo and egg transplants by M. C. Chang, S. Pickworth and P. Dziuk.

HALLER, MARK H. *Eugenics: Hereditarian Attitudes in American Thought*. New Brunswick, N. J.: Rutgers University Press, 1963.

HANEY, BETTE M. and M. W. OLSEN. "Parthenogenesis in premature and newly laid turkey eggs." *Journal of Experimental Zoology*, December 1958.

HASELDEN, KYLE and PHILIP HEFNER, ed. *Changing Man: The Threat and the Promise*. New York: Doubleday, 1969. A varied collection of creative and common essays by theologians and scientists.

HUNT, MORTON. *The Affair: A Portrait of Extramarital Love in Contemporary America*. New York: New American Library, 1970.

HUXLEY, JULIAN. "Eugenics in Evolutionary Perspective." *Perspectives in Biology and Medicine*, Winter 1963.

HOAGLAND, HUDSON and RALPH W. BURHOE, eds. *Evolution and Man's Progress*. New York: Columbia University Press, 1962. A somewhat dated but still very useful collection of essays by James F. Crow, Hermann J. Muller, B. F. Skinner, and others.

JEANNIÈRE, ABEL. *The Anthropology of Sex*. New York: Harper & Row, 1967. Translated from the French, this work takes a valiant step toward a process view of the sexual human. The preface by Daniel Sullivan is very good.

JOYCE, ROBERT and MARY. *New Dynamics in Sexual Love: A Revolutionary Approach to Marriage and Celibacy*. Collegeville, Minn.: St. John's University Press, 1969. An exposition and defense of the celibate or "celebrational" marriage.

KAHN, HERMAN and ANTHONY J. WIENER. *The Year 2000. A Framework for Speculation on the Next Thirty-three Years.* New York: Macmillan, 1967. An authoritative and indispensable guide to considering man's immediate future.

KATZ, JAY. "Experiments on People—What Are the Limits?" Yale University School of Medicine lecture, January 14, 1965.

KLEIN, ISAAC. "Autopsy and Abortion." *Lectures on Medical Ethics 1963–1964.* Yale University School of Medicine. (Mimeographed)

——. "Sterilization, Contraception and Artificial Insemination." *Ibid.*

KOLOBOW, T., W. M. ZAPOL, et al. "Artificial Placenta: Two Days of Total Extrauterine Support of the Isolated Premature Lamb Fetus." *Science,* October 31, 1969.

LEAR, JOHN. "Who Should Govern Medicine?" *Saturday Review,* June 5, 1965.

——. "Do We Need Rules for Experiments on People?" *Saturday Review,* February 5, 1966.

——. "Policing the Consequences of Science." *Saturday Review,* December 2, 1967.

LEDERBERG, JOSHUA. "Humanics and Genetic Engineering." *1970 Britannica Yearbook of Science and the Future,* pp. 81–97.

LEISMER, LUISE. "Trial Marriages for All Couples Urged in State." *The Detroit News,* December 22, 1969. Financial edition.

LEVINE, SEYMOUR. "Sex Differences in the Brain." *Scientific American,* April 1966.

Life. "A turkey that never had a father." April 10, 1956.

——. "Alive in an Artificial Womb." August 28, 1964. Report of Dr. John C. Callaghan's work with lamb fetuses in artificial wombs.

——. "The fantastic drug that creates quintuplets." August 13, 1965.

——. "The control of life." A four-part article, one of the best and among the first to explore the new embryology. September 10, 17, and 23 and October 1, 1965.

——. "Wombs with windows." July 29, 1966.

——. "The challenge to the miracle of life." June 13, 1969. A very important analysis and prognosis for the future of marriage and the interpersonal relationships of men and women; thorough and detailed, for its article length, and well grounded in the latest developments in medicine. By Albert Rosenfeld, former science editor for *Life.* Many of these ideas have been developed more in *The Second Genesis,* by the same writer.

——. "Watching the unborn inside the womb." July 25, 1969.

Lo Bello, Nino. "Coming Soon: World's First Artificial Baby." *Science Digest,* February 1966. Sober report of the controversial work of

Dr. Daniele Petrucci with the artificial womb and human development.

LOMAX, LOUIS E. "Sperm Bank—the brutal truth we dare not face." *Pageant,* January 1958.

LOVELOCK, J. E. and M. H. BISHOP. "Prevention of Freezing Damage to Living Cells by DMSO." *Nature,* 183:1394, 1959.

LURIA, SALVADOR E. "Directed Genetic Change: Perspectives from Molecular Genetics." See Sonneborn below.

LUYET, B. J. *Survival of Cells, Tissues and Organisms in Freezing and Drying.* New York: Hafner, 1952.

——, and GEHENIO, P. M. *Life and Death at Low Temperatures.* Normandy, Mo.: Biodynamica, 1940. Dated, early study.

MALINOWSKI, BRONISLAW. *The Sexual Life of Savages in North-Western Melanesia.* London: Routledge and Kegan Paul, 1932.

——. *Sex, Culture and Myth.* New York: Harcourt, Brace and World, 1962.

MASLOW, ABRAHAM H. *Toward a Psychology of Being.* Princeton: Van Nostrand, 1962.

——. "Love in Self-Actualizing People." See De Martino above.

MASTERS, R. E. L. *Forbidden Sexual Behavior and Morality.* New York: Julian Press and Lancer Books, 1962. Comprehensive and informative survey of various sexual customs in a variety of contemporary and ancient societies; helpful in appreciating our present evolving concepts.

MCLUHAN, MARSHALL and GEORGE B. LEONARD. "The Future of Sex." *Look,* July 25, 1967. A provocative, controversial and creative appraisal of current trends in American and Western man's life. Basic reading.

MEAD, MARGARET. *Male and Female: A Study of the Sexes in a Changing World.* New York: William Morrow, 1949. A classic reference on marriage and sexuality in contemporary America.

——. *Growing Up in New Guinea.* New York: William Morrow, 1930. 1968 revised edition published by Dell.

——. *Sex and Temperament in Three Primitive Societies.* New York: William Morrow, 1935. 1963 edition by Dell. All of Margaret Mead's books are extremely useful in appreciating the varieties of sexual images and patterns evolved in the history of man.

——. "Why Just 'Living Together' Won't Work." *Redbook,* April 1968. The author takes a second and critical look at her proposal for trial marriages.

MESSENGER, ERNEST C. *Evolution and Theology.* London: Burns Oates and Washbourne, 1931.

——. *Two in One Flesh*. New York: Newman, 1956.

——. ed. *Theology and Evolution*. London: Sands, 1949.

MILLER, NEAL E. "Physiological and cultural determinants of behavior." In *The Scientific Endeavor*. New York: Rockefeller University Press, 1965.

MINSKY, MARVIN L. "Artificial Intelligence." *Scientific American*, September 1966.

MONEY, JOHN, ed. *Sex Research: New Developments*. New York: Holt, Rinehart and Winston, 1965. A dozen valuable essays surveying new findings in psychobiology: Phoenix on hormones and sexual behavior; Pfaff on cerebral implantations, MacLean on psychosexual functions of the brain and Money on psychosexual differentiation.

MONTAGU, ASHLEY. *Sex, Man and Society*. New York: Putnam's Sons, 1969. A motley collection of previously published esoteric papers, but some worthwhile perceptive comments on the contraceptive pill, sex education, and the sexual revolution.

——. ed. *Marriage: Past and Present. A Debate Between Robert Briffault and Bronislaw Malinowski*. Boston: Porter Sargent, 1956.

MOORE, KEITH L. "The Vulnerable Embryo: Causes of Malformation in Man." *Manitoba Medical Review*, June–July 1963.

MUELLER, GERHARD. "Toward Ending the Double-Standard of Sexual Morality." *Journal of Offender Therapy*, 8:1, 1964.

MULLER, HERMANN J. "Human Evolution by Voluntary Choice of Germ Plasm." *Science*, September 8, 1961. Muller's classic exposition of a eugenics program.

——. "Should We Weaken or Strengthen Our Genetic Heritage?" *Daedalus*, Summer 1961.

——. "Genetic progress by voluntarily conducted germinal choice." See Wolstenholme below.

——. "Means and Aims in Human Betterment." See Sonneborn below.

NELSON, J. ROBERT. "Morals and Medicine." *The Commonweal*. December 13, 1968. An important summary of the lack of theological guidelines for a new ethics of medical research.

NEMECEK, OTTOKAR. *Virginity: Pre-nuptial Rites and Rituals*. New York: Philosophical Library, 1958.

NEUBECK, GERHARD, ed. *Extramarital Relations*. Englewood Cliffs, N.J.: Prentice-Hall, 1969. A broad spectrum of essays which should prove helpful in appraising the impact of the new embryology on fidelity in marriage.

Newsweek. "Upsetting the Odds." October 14, 1968.

——. "How Are Your Genes?" February 10, 1969.

——. "World of the Unborn." February 26, 1968.

A Selected Bibliography with Some Annotations   273

——. "Monitoring the Unborn." October 21, 1968.

——. "Anatomy of Violence." January 12, 1970. Recent research with the XYY chromosome abberation.

OLSEN, M. W. "Fatherless Turkey." *Journal of the American Veterinary Medical Association,* January 1, 1959.

——. "Nine-year summary of parthenogenesis in turkeys." *Proceedings of the Society for Experimental Biology and Medicine,* 105, 1960.

——. "Performance Record of a Parthenogenetic Turkey Male." *Science,* 132, December 2, 1960.

Ortho Research Foundation, Hartman Library Bibliography. Number 3, September 1968. "Selected Bibliography on Induced Superovulation in Livestock and Laboratory Animals, 1939–1968."

OSBORN, FREDERICK. *The Future of Human Heredity. An Introduction to Eugenics in Modern Society.* New York: Weybright and Talley, 1968. A terse and valuable historical survey of eugenics and its practical problems for application in man's future.

PACKARD, VANCE. *The Sexual Wilderness. The Contemporary Upheaval in Male-Female Relationships.* New York: David McKay, 1968. Thorough and indispensable analysis of trends and present situation.

PARKES, A. S. *Sex, Science and Society.* Chester Springs, Pa.: Dufour, 1968. Contains a variety of talks on topics related to sexuality and research on this topic.

Pathways to Conception: The Role of the Cervix and the Oviduct in Reproduction. A symposium sponsored by the Department of Gynecology and Obstetrics, Wayne State University, November 13 and 14, 1969. A panel discussion of "Test Tube Babies—Fact or Fancy," plus formal papers by Dr. Philip Dziuk on "Egg Transfer," Dr. R. L. Brinster on "In Vitro Culture of the Embryo," and Dr. S. J. Behrman on "Immunological Phenomena of Reproduction" are among the less technical contributions to this high-level symposium.

PERRY, ENOS J., ed. *The Artificial Insemination of Farm Animals.* New Brunswick, N.J.: Rutgers University Press, 1952. An opening chapter by the editor contains a comprehensive historical survey of the technique.

PETRUCCI, DANIELE. "Producing Transplantable Human Tissue in the Laboratory." *Discovery,* July 1961.

PINCUS, GREGORY. *The Control of Fertility.* New York: Academic Press, 1965.

——. "The Comparative Behavior of Mammalian Eggs in Vivo and in Vitro. IV. The Development of Fertilized and Artificially Activated

Rabbit Eggs." *Journal of Experimental Zoology,* 82:85–129, 1939.
———. "Observations on the living eggs of the rabbit." *Proceedings of the Royal Society (Biology),* 107:132, 1930.
———, and H. SHAPIRO. "The Comparative Behavior of Mammalian Eggs in Vivo and in Vitro. VII. Further Studies of Activation of Rabbit Eggs." *Proceedings of the American Philosophical Society,* 83:631–647, 1940.
———, and B. SAUNDERS. "Maturation of Human Follicular Ova in Vitro." *Anatomical Record,* 75:537, 1939.
PIROVSKY, BERNARD, et al. "Production of Immune Tolerance in Humans." *Nature,* April 20, 1968. An early report of experiments with the suppression of the immune response in newborn humans by injection of foreign blood proteins.
Psychology Today. July 1969. A nicely rounded issue with informative popular essays by Gordon Bermant on animal sexual behavior, Frank Beach on psychosexuality in animals and humans, Jerome Kagan on social roles of the male and female and an interview with Johnson and Masters, authors of *The Human Sexual Response.*
REISS, IRA L. *Premarital Sexual Standards in America.* Glencoe, Ill.: The Free Press, 1960.
———. "How and Why America's Sex Standards Are Changing." *Trans-Action,* March 1968.
Research in Reproduction. "Antibodies and Conception." January 1969.
———. "Gonocytes and sex determination." January 1969.
———. "Fertilization of mammalian eggs *in vitro.*" March 1969.
———. "Hormonal control of the implantation of embryos." May 1969.
———. "The origin of sex chromosome imbalance in man." May 1969.
———. "Reproduction in African animals." May 1969.
This journal, published bimonthly, is edited by Dr. R. G. Edwards, of the Physiological Laboratory at Cambridge. Each four-page issue contains a wealth of information on the latest advances in reproductive research, often in reports by the pioneer scientists themselves.
RIMMER, ROBERT H. *The Harrad Experiment.* Los Angeles: Sherbourne Press, 1966. A serious and fascinating utopian novel focusing on the psychological and emotional developments of three young couples involved in an experimental college program designed to promote the evolution of new patterns of male-female relationship and a new sexual ethics.
———. *Proposition 31.* New York: New American Library, 1968. A sequel to *The Harrad Experiment* focusing again on psychological

and emotional developments applied to the evolution of group marriage among the novel's characters.

——. *The Rebellion of Yale Marratt.* New York: Avon Books, 1967. Explores the possibilities of a co-marital or bigamous relationship between consenting adults.

ROBIN, EUGENE D. "Rapid Scientific Advances Bring New Ethical Questions." *Journal of the American Medical Association,* August 24, 1964.

ROCK, JOHN and M. F. MENKIN. "In Vitro Fertilization and Cleavage of Human Ovarian Eggs." *Science,* 100:105, 1944.

——. "In Vitro Fertilization and Cleavage of Human Ovarian Eggs." *American Journal of Obstetrics and Gynecology,* 55:440, 1948.

ROCK, JOHN and A. T. HERTIG. "The Human Conceptus During the First Two Weeks of Gestation." *American Journal of Obstetrics and Gynecology,* 55:6, 1948.

RORVIK, DAVID M. "The Brave New World of the Unborn." *Look,* November 4, 1969. A report on semidelivery and Dr. Asensio's work.

——. "Making Men and Women Without Men and Women." *Esquire,* April 1969. A sensational approach but with some interesting contents and quotes on cloning and genetic engineering.

——. "Artificial Inovulation." *McCall's.* May 1969.

——. "Cloning: asexual human reproduction." *Science Digest.* November 1969.

ROSENFELD, ALBERT. *The Second Genesis: The Coming Control of Life.* Englewood Cliffs, N.J.: Prentice Hall, 1969. A major survey of the biological revolution by the former science editor of *Life.*

——. "The Futuristic Riddle of Reproduction." *Coronet,* February 1959.

——. "Drama of Life Before Birth." *Life,* April 30, 1965.

——. "The New Man: What Will He Be Like?" *Life,* October 1, 1965.

——. "A Laboratory Study of Sexual Behavior." *Life,* April 22, 1966.

——. "Science Is Where the Action Is." *Life,* July 29, 1966.

——. "The Scientists' Findings: More Sex, Less Promiscuity." *Life,* May 31, 1968.

ROSTAND, JEAN. *Can Man Be Modified?* New York: Basic Books, 1959. A classic, but very dated.

——, and ALBERT DELAUNAY, ed. *Man Of Tomorrow,* Volume 8 of the *Encyclopedia of the Life Sciences.* Garden City: Doubleday, 1966.

ROUGEMONT, DENIS DE. *Love in the Western World.* New York: Pantheon, 1940. The classic documentation and analysis in this field.

276   *Utopian Motherhood*

Roy, Rustum and Della. *Honest Sex. A Revolutionary Sex Ethics By and For Concerned Christians.* New York: New American Library, 1968. A pioneering ethical study basic for any serious approach to the evolution of human sexuality. Particularly good in its exposition of co-marital sexual expressions.

Rubin, Bernard. "Psychological Aspects of Human Artificial Insemination." *Archives of General Psychiatry,* August 1965.

Scheinfeld, Amram. *Women and Men.* New York: Harcourt, Brace, 1943. An older work, but still valuable as an overview of the complexities of human sexuality.

Schur, Edwin M., ed. *The Family and the Sexual Revolution.* Bloomington: Indiana University Press, 1964. Essays on changing sex standards include discussion of trial marriage, premarital sex, the double standard, sex and temperament, the kibbutz, and forms of birth control.

Science. "Human Experimentation: New York Verdict Affirms Patient's Rights." February 11, 1966.

Science News. "Mice in a Test Tube." December 14, 1968.

———. "Conception: a controversy." September 14, 1968. Exchange of views between Dr. Paul Ramsey, theologian, and Dr. Bent G. Böving, embryologist.

Scientific Research. "Tree Produced by Tissue Culture." April 28, 1969.

Shainess, Natalie. "Comment on 'Marriage of the Future.'" *Marriage,* January 1969. A response and critique of the "celebrational" or celibate marriage proposed by Robert and Mary Joyce.

Sherman, Jerome K. "Improved Method for Frozen Storage of Human Spermatozoa." A paper delivered at the annual meeting of the Federation of American Societies for Experimental Biology, April 1962.

Shettles, Landrum B. "Parthenogenetic Cleavage of the Human Ovum." *Bulletin of the Sloane Hospital for Women,* June 1957.

———. "Observations on Human Follicular and Tubal Ova." *American Journal of Obstetrics and Gynecology,* 66:235, 1953.

———. "Corona Radiata Cells of the Zona Pellucida of Living Human Ova." *Fertility and Sterility,* 9:167, 1958.

———. "Morula Stage of Human Ovum Development in Vitro." *Fertility and Sterility,* 6:287, 1955.

———. and David M. Rorvik. "Choose Your Baby's Sex." *Look.* April 21, 1970. A major popular article detailing a simple safe do-it-at-home method for predetermining the sex of offspring.

SMITH, A. U. *Biological Effects of Freezing and Supercooling*. Baltimore: Williams and Wilkins, 1961.

——. "Behavior of Fertilized Rabbit Eggs Exposed to Glycerol and to Low Temperatures." *Nature*, 170:374–375, 1952.

SONNEBORN, TRACY M., ed. *The Control of Human Heredity and Evolution*. New York: Macmillan, 1965.

SPENCER, STEVEN M. "The Birth Control Revolution." *Saturday Evening Post*, January 15, 1966.

——. "The Pill That Helps You Remember." *Saturday Evening Post*, September 24, 1966.

STEWARD, F. C. "The Control of Growth in Plant Cells." *Scientific American*, October 1963.

STOCK, ROBERT W. "Will the Baby Be Normal?" The *New York Times Magazine*, March 23, 1969. A fine survey of genetic counseling.

——. "The XYY and the Criminal." *New York Times Magazine*. October 20, 1968.

TAYLOR, GORDON RATTRAY. *The Biological Time Bomb*. New York: World Publishing Co., 1968. The first and still useful survey of the whole biological revolution by a leading British science writer; more oriented toward legal and economic implications than value issues.

The Lancet. "Parthenogenesis in Mammals?" November 5, 1955.

Time. "Parthenogenesis?" November 21, 1955.

——. "Sex to Order?" October 7, 1957.

——. "The Secret of Life." July 14, 1958.

——. "Survival of the Unfit?" June 16, 1958.

——. "Sex by Sedimentation." July 17, 1964.

——. "The Age of Alloplasty." January 1, 1965.

——. "Transfusions in the Womb." January 15, 1965.

——. "'Vaccinating' against Rh." October 8, 1965.

——. "The Riddle of A.I." February 25, 1966.

——. "A Molecule for Memory?" January 7, 1966.

——. "Pills to Keep Women Young." April 1, 1966.

——. "How to Detect a Faulty Gene." September 12, 1969.

——. "Is Intercourse a Factor?" November 14, 1969.

——. "An Elegant Triumph." December 5, 1969. Isolation of the lactose gene.

UNGAR, GEORGES. "Chemical Transfer of Learned Behavior." A paper delivered at the December 29, 1967, meeting of the American Association for the Advancement of Science.

VAN HORNE, H. "Are We the Last Married Generation?" *McCall's*. May 1969. A substantial critique of "trial marriages," but lacking in any positive or creative proposal for resolving the present trends.

WARSHOFSKY, FRED. *The Control of Life: The 21st Century.* New York: Viking Press, 1970.

WENDT, HERBERT. *The Sex Life of the Animals.* New York: Simon and Schuster, 1965. Excellent popular history of man's growing knowledge of sexuality.

WHITMAN, ARDIS. "Is Marriage Still Sacred?" *Redbook,* February 1967.

WHITNEY, LEON F. "The Successful Transfer of Ovaries between Dogs of Different Breeds." *Science,* May 24, 1946.

———. "Ovarian Transplantation in Dogs." *Veterinary Medicine,* January 1947.

WILLETT, E. L. "Egg Transfer and Superovulation in Farm Animals." *Iowa State Journal of Science,* September 1953.

———, et al. "Three Successful Transplantations of Fertilized Bovine Eggs." *Journal of Dairy Science,* May 1953.

WILLIAMS, GLANVILLE. *The Sanctity of Life and the Criminal Law.* New York: Alfred A. Knopf, 1967.

WINCHESTER, JAMES H. "Babies Without Fathers!" *Sunday Mirror Magazine,* January 29, 1956.

WINICK, CHARLES. *The New People. Desexualization in American Life.* New York: Pegasus, 1968. A controversial and seriously disturbing analysis of the supposed American movement toward a unisexual society.

WITSCHI, EMIL. "Sex Reversal in Animals and in Man." *American Scientist,* September 1960.

WOLSTENHOLME, GORDON, ed. *Man and His Future.* Boston: Little Brown, 1963. Contains provocative essays by Julian Huxley, J. B. S. Haldane, Gregory Pincus, A. S. Parkes, Alex Comfort, Joshua Lederberg, Hudson Hoagland, et al.

YANAGIMACHI, R. and M. C. CHANG. "In Vitro Fertilization of Golden Hamster Ova." *Journal of Experimental Zoology,* 156:361, 1964.

YOUNG, LOUISE B., ed. *Evolution of Man.* New York: Oxford University Press, 1970. Very informative and varied views offered in a well-balanced collection of short essays on such topics as birth and death control, selective breeding, new techniques of positive eugenics, utopias, and antiutopias.

YUNCKER, B., "Baby Bubble." *Ladies Home Journal,* September 1969. An excellent review of research with Heyns's decompression technique for superoxygenating babies prenatally.